The Strategic Shape of the World

The Strategic Shape of the World

Proceedings of MEA–IISS Foreign Policy Dialogue

Edited by
Amit Dasgupta

⑤SAGE Los Angeles • London • New Delhi • Singapore
www.sagepublications.com

First published in 2008 by

SAGE Publications India Pvt Ltd
B1/I-1 Mohan Cooperative Industrial Area
Mathura Road, New Delhi 110 044, India
www.sagepub.in

SAGE Publications Inc
2455 Teller Road
Thousand Oaks, California 91320, USA

SAGE Publications Ltd
1 Oliver's Yard, 55 City Road
London EC1Y 1SP, United Kingdom

SAGE Publications Asia-Pacific Pte Ltd
33 Pekin Street, #02-01 Far East Square
Singapore 048763

Published by Vivek Mehra for SAGE Publications India Pvt Ltd, typeset in 10.5/12.5 Aldine401 BT by Star Compugraphics and printed at Chaman Enterprises, New Delhi.

Library of Congress Cataloging-in-Publication Data Available.

ISBN: 978-81-7829-936-5 (Pb)

The SAGE Team: Rekha Natarajan, Rachna Sinha and Trinankur Banerjee

Contents

Preface

A unique feature of the Indian effort in public diplomacy is that the Ministry of External Affairs (MEA) not only engages the civil society and academia within India but also abroad. Such a dialogue allows the Ministry the opportunity to communicate the compulsions behind our foreign policy concerns and to note how our foreign policy initiatives are perceived, both within India and abroad. Among the efforts that have been initiated outside India is a Foreign Policy Dialogue with the influential international think-tank, the International Institute for Strategic Studies (IISS), London. The first Dialogue took place in February 2007, London (see Appendix 2 for the programme details).

The Second MEA–IISS Foreign Policy Dialogue on 'The Strategic Shape of the World' was held in New Delhi, India on 13 December 2007. The event was organized by the Public Diplomacy Division, MEA with the support of the IISS (see Appendix 1 for the programme details).

The theme of the Dialogue 'The Strategic Shape of the World' was identified because of the felt need to examine and analyze key issues of global importance and how these could affect international relations and strategic alliances. These issues included concerns regarding the deepening of globalization, international security and stability, transnational economic dimensions, energy security, and climate change, among others. The Dialogue also provided a unique opportunity to discuss India's role in the global order and its impact on the regional and global environment.

The event carried forward the dynamic partnership between MEA and the IISS. It brought together representatives of governments, universities, policy institutions, research centres, the media and the intelligentsia from India and the IISS. It opened with the keynote address by the Indian Foreign Secretary. His Excellency S.K. Singh, Governor of Rajasthan, agreed to deliver a special address prior to the post-lunch session.

The Dialogue was divided into three plenary sessions preceded by the inaugural session. The three sessions titled 'The Strategic Shape of the World', 'International Terrorism' and 'Energy Security' provided a public platform for interaction between the experts of national and international repute. Thus, yet again, the Dialogue proved to be a success in promoting the key objective of bringing together policy makers and experts as well as promoting a vigorous debate on foreign policy issues.

The Public Diplomacy Division of the MEA is grateful for the support of IISS, London and the participating delegates for making the event a meaningful experience.

Amit Dasgupta
New Delhi, September 2008

Inaugural Session

Speakers
CHAIRMAN: Amit Dasgupta, Joint Secretary, Public Diplomacy, MEA
ARIF S. KHAN: Additional Secretary, Public Diplomacy, MEA
SHIVSHANKAR MENON: Foreign Secretary of India
PATRICK CRONIN: Lead Speaker and Director of Studies, IISS

Chairman: Good morning everyone. Without further ado, I would like to request the Additional Secretary in the Ministry of External Affairs (MEA), Arif S. Khan, to say a few words and to invite the Foreign Secretary to inaugurate the Conference.

Arif S. Khan: Excellencies, distinguished colleagues, ladies and gentlemen—

It is a privilege for us to host this particular event after attending the same in England earlier this year. I am happy to welcome Dr Cronin and his colleagues from The International Institute for Strategic Studies (IISS), London. I am also honoured that the Foreign Secretary, Shivshankar Menon, is with us today, despite his busy schedule. His presence will set the tone for this seminar. I take this opportunity to invite the Foreign Secretary.

Shivshankar Menon: Thank you Arif, Dr Cronin, Amit, ladies and gentlemen.

I am delighted to inaugurate this Foreign Policy Dialogue with the IISS, which was initiated earlier this year. The IISS has played a significant role in shaping international discourse on strategic issues and I am happy to note that it is now concentrating on Asia. In April, IISS will organize a major conference on India. Such occasions offer a useful opportunity for experts and scholars to exchange and analyze views on the series of strategic issues that confront us.

Today's one-day dialogue will engage on three subjects: the strategic shape of the world, international terrorism and energy security. All three are crucial to whether or not we in India achieve the basic goal we have set for our foreign policy: of enabling the transformation of India. In fact, today, more than ever, the outside world will affect India's future. Measured by any criterion, the proportion of gross domestic product (GDP) linked to foreign trade, the role of foreign investment and technology, and India's need for energy, markets and raw materials, we are more connected to the rest of the world than ever before. The strategic shape of the world and the issues that you will consider are therefore central to our concerns.

The strategic shape of the world

It is sometimes said that what you see depends on where you stand. Looking at the world from India, it often seems that we are witness to the collapse of the Westphalian state system and a redistribution in the global balance of power, leading to the rise of major new powers and forces. Our shorthand for this phenomenon is the rather inadequate word, 'globalization'. Equally, the twin processes of globalization and economic interdependence have resulted in a situation where Cold War concepts like containment have very little relevance. The interdependence brought about by globalization puts limits beyond which tensions among the major powers cannot escalate. What seems likely, and is in fact happening, is that major powers come together to form coalitions to deal with issues where they have a convergence of interests, despite differences in the broader approach. In other words, what we see is the emergence of a global order marked by the preponderance of several major powers, with minimal likelihood of direct conflict amongst these powers. The result is a de-hyphenation of relationships with each other, of each major power engaging with all the others in a situation that might perhaps be described as 'general un-alignment'.

We see the evolving situation as one in which there is an opportunity for India. As a developing country, the consistent objective of our foreign policy has been and remains poverty eradication and rapid and inclusive economic development. If we are to eradicate mass poverty by 2020, we need to keep growing our economy at 8–10 per cent each year. This requires a peaceful and supportive global environment in general and a peaceful periphery in particular. We have therefore attempted since Independence to expand India's strategic space, thus strengthening our strategic autonomy. In practice, this has meant the increasing autonomy of decision-making for India on issues that are of importance to us—what our first Prime Minister Jawaharlal Nehru used to call our 'enlightened self-interest'.

Today, the international situation has made possible the rapid development of our relationships with each of the major powers, and this is apparent in developments over the last few years. Equally important have been two other necessary conditions for this rapid development which have given India space to work in: India's rapid economic and social transformation. As a result of 25 years of 6 per cent growth and our reforms since 1991, India is today in a position to engage with the world in an unprecedented manner. Our engagement with the global economy is growing rapidly, with trade in goods and services now exceeding US dollars 330 billion. Our needs from the world have changed, as has our capability. India can do and consider things that we could not do or consider 20 years ago. This is reflected in how India perceives its own future, its ties in its neighbourhood and its approach towards the larger international order. The second necessary condition which has obtained to a greater or lesser extent is our attempt to build a peaceful periphery within which India's transformation can take place.

Paradoxically, some of the same forces of globalization—the evolution of technology, the mobility of capital, and so on—which have led to the decline or collapse of the Westphalian state order are also the source of our greatest dangers. Our major threats today are from non-state actors, from transboundary effects of the collapse of the state system, or, at least, of its inadequacy. (Parenthetically, the doctrine of absolute sovereignty created by the strong European states and rulers in earlier centuries is now the last defence of the weak against the strong.)

Looking ahead, the real factors of risk that threaten systemic stability come from larger global issues like terrorism, energy security and environmental and climate change. With globalization and the spread of technology, the threats have also globalized and now span across borders. These are issues that will directly impact India's ability to grow and expand its strategic autonomy. It is also obvious that no single country can deal with these issues alone. The issues require global solutions. Hence the importance of what you will discuss about the strategic shape of the world.

International terrorism

Among these global threats, international terrorism remains a major threat to peace and stability. We in India have directly suffered the consequences of the linkages and relationships among terrorist organizations, support structures and funding mechanisms, centred upon our immediate neighbourhood, and transcending national borders. Any compromise with such forces, howsoever pragmatic or opportune it might appear momentarily, only encourages the forces responsible for terrorism. Large areas abutting India to the West have seen the collapse of state structures and the absence of governance or the writ of the state, with the emergence of multiple centres of power. The results, in the form of terrorism, extremism and radicalism, are felt by us in India.

Energy security

As for energy security, this is one issue which combines an ethical challenge to all societies with an opportunity to provide the energy so necessary for development. For India, clean, convenient and affordable energy is a critical necessity if we are to improve the lives of our people. Today, India's per capita energy consumption is less than a third of the global average (only 500 kilogramme of oil equivalent [kgoe] compared to a global average of nearly 1,800 kgoe). For India, a rapid increase in energy use per capita is imperative to realize its national development goals.

Global warming and climate change require all societies to work together. While the major responsibility for the accumulation of greenhouse gasses (GHGs) in the atmosphere lies with the developed countries, its adverse effects are felt most severely by developing countries like India. When we speak of 'shared responsibility', it must include the international community's shared responsibility to ensure the right to development of the developing countries. Development is the best form of adaptation to climate change.

At the South Asian Association for Regional Cooperation (SAARC) Council of Ministers last week in Delhi, a declaration was adopted which noted that the way forward must include:

1. Adequate resources to tackle climate change without detracting from development funds.
2. Effective access to and funding assistance for the transfer of environment-friendly technologies and for adaptation.
3. Binding GHG emission reduction commitments by developed countries with effective time frames.
4. Equitable burden-sharing.

Equally, the transfer and access to clean technologies by developing countries, as global public goods on the lines of what was done for retrovirals to fight AIDS, is essential to effectively limit future GHG emissions. The Intellectual Property Rights (IPR) regime should include collaborative research and development (R&D) and the sharing of the resulting IPRs.

India will work constructively with the global community to find solutions which do not constrain development. We are determined not to allow our per capita emissions of GHGs to exceed those of developed countries. For all these reasons, the subjects you are about to consider are central to our concerns in India. I look forward to the results of your deliberations and wish you a very successful dialogue.

Thank you very much.

Chairman: May I now invite Dr Patrick Cronin to say a few words?

Patrick Cronin: Mr Foreign Secretary, I think you have made such a clear articulation of the big challenges that we face in this dialogue that we are all stymied from asking you the detailed questions that are required for looking for solutions. But that is what the purpose of this dialogue is all about. You honour us a great deal, Sir, with your remarks here today. This is something that has been near and dear to the heart of Arif Khan and others from the IISS and the MEA. Your remarks here while inaugurating this session of the MEA and the IISS will help make sure that this is a dialogue that carries forth into the new year (2008) and the years beyond. We are very grateful to you, Sir, for this and for your strategic leadership around the world as well. You have spoken at the IISS before. We have members in 100 countries. India may be a rising great power but its Foreign Secretary is already a great Foreign Secretary. Thank you, again.

[*Inaugural Session Ends*]

Session I

The Strategic Shape of the World

Speakers
Chairman: Amit Dasgupta, Joint Secretary, Public Diplomacy, MEA
Patrick Cronin: Lead Speaker and Director of Studies, IISS
Ummu Salma Bava: Indian Lead Speaker (Professor of European Studies, Jawaharlal Nehru University, New Delhi)
Sir Michael Quinlan: IISS Discussant
Dr Manpreet Sethi: Indian Discussant
R.B. Mardhekar: Additional Secretary, MEA
Arundhati Ghose: Former Ambassador
Ishrat Aziz: Former Ambassador
Indraneel Banerjee: Researcher
Meena Singh Roy: Institute for Defence Studies and Analyses, New Delhi
A.S. Kalkat: Lieutenant General (Retired)
B. Raman: Former Additional Secretary

Chairman: We will now commence the substantive sessions of the Foreign Policy Dialogue. To briefly recap, this Dialogue was formally initiated earlier this year when the IISS hosted the Indian delegation in London. It is now our turn to be the hosts. The first business session, as the Foreign Secretary mentioned earlier this morning, is titled 'The Strategic Shape of the World'.

The rules of the game are, as you can see: there are four speakers; the deliberations will be opened by Dr Patrick Cronin, who will be the lead speaker from the IISS; there is also a discussant; so, each side has a lead speaker and a discussant. We are hoping that everyone would stick to the time limit, which is 15 minutes for the lead speaker and 10 minutes for the discussant. Thereafter, we would open the floor for discussions.

The team leader of the IISS is Dr Patrick Cronin, the Director of Studies. He joined IISS in September 2005 and was earlier with the Center for Strategic and International Studies in Washington. Dr Cronin, the floor is yours.

Patrick Cronin: Thank you very much. I think I am going to speak sitting down because there is a laptop up at the podium and my notes will slide off onto the floor, and that would be very awkward.

I am fortunately accompanied by a very distinguished team of experts from the IISS, not least among whom is Sir Michael Quinlan, who will be on the same panel as the discussant. So, I feel that I can afford to be somewhat superficial in the question that I have been given this morning, which inevitably invites superficiality because the strategic shape of the world is a question for the ages. It is probably not the shape of the world that has changed as much as our perceptions and our concepts about it. I think the Foreign Secretary struck exactly the right notes. When I put forward three general propositions of what does one even begin to think about the strategic shape of the world, I believe my three propositions would be very much in line with what the Foreign Secretary said.

The deepening globalization of the world, as experienced, has not been without costs as well as benefits. One of the costs is that the balance of power has gone from the bipolar world, especially those of us in the West and us in America changing the concept, to not so much the multi-polar world yet but to what our Director-General at the IISS likes to call a non-polar world, a world in which not even the United States (US), still the dominant power, is able to galvanize and organize that coherence that the Foreign Secretary said needs to be converged if we are to have coherent policies to grapple with the challenges.

The second proposition is that most of those challenges, again to echo the Foreign Secretary, are transnational, transboundary and global in nature—terrorism and the hybrid conflicts that we are experiencing which are no longer just civil wars or inter-state. Proliferation, energy security, global pandemics, climate change and other issues—so many of the challenges that we face are indeed these global transnational challenges that the state system and the major powers are not really equipped or have not moved to transition to grapple with.

The third proposition is to emphasize that these are not all threats in terms of the old paradigm either—that there are equal number of opportunities that are out there waiting for a younger generation to seize the opportunity to do some unprecedented things. I want to just touch on those very briefly just to remind us all that even though at the strategic studies business we focus on threats and challenges in a sort of a worst-case scenario way, we also need to think about what is the purpose of resolving these challenges, whether we are trying to achieve what I have referred to back in the US as the 'long peace', not the 'long war'.

What are the five challenges that jump out at me from the agenda of the IISS as well as from my own work, especially on US policy around the world? I think of the five that first come to mind. Maybe I will emphasize the first three the most. I start with terrorism because, indeed, we seemed to have entered a new era on 11 September 2001. My distinguished colleague Nigel Inkster is an expert on these issues. He is going to talk at length later today. I want to just skim the surface of a couple of points here.

We know that terrorism is a tactic. We know that it is not new. So, why is this a new era? I think that one of the reasons why the strategic studies community has been preoccupied with it is because of globalization. The effects of terrorism today can truly be global and transnational and more lethal; we are seemingly in this open-ended threat. In a way terrorism was much more confined in the past. At least that is the fear—that extremists can in a globalized world appeal to a much larger audience. Ultimately, it sets up this global contest—this transnational global contest—in which we are each trying to mobilize opinion, call it hearts and minds, or the opinion of the communities, especially Muslim communities these days, as these extremists are often distorting Islam in particular as part of a narrative in which the bottom line is, 'The West is repressing and trying to keep down Muslims'. This narrative is a very subtle, complex mobilization contest.

Unfortunately, the response, especially from the US, my own country, in the last six years has been to reach out and use the most available tool, which has been the military instrument of power. That response in itself has again generated this paradigm. It has, this global war on terror has, even more deeply embedded the paradigm, that of the clash of civilizations. This type of philosophy, this type of thinking, further joins the game, if you will, and is making this seem like a new paradigm. So, you have to be very careful obviously on how you respond to the challenges that you identify. Again Nigel Inkster will be disaggregating rather than conflating some of these challenges. Of course, that is indeed part of what has to happen if we are to address them.

'Hybrid conflict' is now the latest phrase that is gaining favour back in Washington. There have been other terms that have been used—asymmetric warfare, irregular warfare, counter-insurgency, etc.—but 'hybrid warfare' better captures some of the complexity again that we are not seeing a simple opponent on the battle field, we are seeing a series of opponents: foreign fighters, indigenous insurgents fighting for national self identity, border skirmishes or interventions from neighbours who want to meddle and intervene in these incidents. Nothing new in the course of history or international affairs, but again with globalization, these forces are more empowered, enabled to deploy farther afield than ever.

In fact, I was in Saudi Arabia a few days ago meeting with a senior official, talking about the time of the Haj and how sensitive a time it is when a young man going on the Haj is inspired—and unfortunately, some are receptive—to being recruited because they are convinced by this narrative that they are fed that this is an honourable thing to do, to go out and kill people, even innocent civilians, because there is this higher calling. That is a very complex challenge that defies any instrument of power and hence, it is bleeding over into these insurgencies and the local conflicts. So, we have got terrorism now blending in with insurgency and this is a challenge not just in Iraq and Afghanistan but well known even to South Asia and also throughout the Middle East, Africa and elsewhere.

Let me just touch very briefly on Iraq to say that there we have seen in the past year, if we can just isolate the shape of the strategic thinking for the past year, the so-called surge that started in February 2007. This resulted from the lapse of a strategy in the recognition of that last December in Washington when the administration was faced with the Iraq Study Group Report, the Bipartisan Committee Report. They put in place a change of tactics militarily led by General David Patreas, who, as you may know, was working with the Iraqi security forces as they were becoming more numerous and capable. But essentially the tactic was to stop focusing on just killing insurgents—that is an endless no-win game—and start focusing on protecting the population and then making sure that behind that military surge there is a political surge, a surge of development, a surge of employment, a surge of opportunity. Easier said than done! That is still one of the shortcomings. The other shortcoming in Iraq remains the question mark around the real desire for unity on the part of the Iraqi government. These are the question marks that will determine the relative success or failure of what happens in a fairly messy Iraq over the next few years.

I think the important point for the US is that however it intervened and however that is viewed, many in the region and beyond stand to be deeply affected by how it leaves Iraq. That is why even the democratic candidates who are running for President in the US election this year are now insisting that they should not be given a due date for withdrawing from Iraq. There is recognition in both parties, especially the Democratic Party, which may very well come to power, that Iraq requires a great deal of care and that it will take some time. So too in Afghanistan as the process of Afghanization has to proceed, more Afghans have to be in charge, and it has to be more than 'Kabul'. These are challenges that countries like Britain, the US and India, and those comprising the North Atlantic Treaty Organization (NATO), who are playing key roles through developmental projects and other means with the objective of ensuring that Afghanistan does not lapse yet again into a failed state, face.

A third challenge for our time, one that Sir Michael Quinlan is going to, I think, address in his comments is nuclear proliferation. Further down the line, we can talk about the proliferation of the life sciences, especially biological warfare. But right now we are still living with the fear of a second nuclear age—that we may be on the cusp of a second age of nuclear proliferation. Next year it will be 50 years since the IISS started. For the first decade it was focused almost exclusively on how to prevent nuclear war, how to stop nuclear proliferation. This has been one of the four issues (concerns) for the institute over all the five decades of its being and it is not going away as we find out today. Globalization again has left its mark on this problem of proliferation. The A.Q. Khan network, which was essentially an amorphous international network wheeling and dealing in the nuclear black market, is exactly the case in point since it globalized proliferation.

That is how globalization has shifted this proliferation problem in the 21st century. We also have the fact again that by the proliferation of regional and local powers, more countries are seeking nuclear weapons as an insurance policy, a guarantee of protection.

I should mention again, just going back now to US policy, the latest National Intelligence Estimate (NIE) which said that Iran is believed to have halted its nuclear weapons programme that it had been dissembling about and pursuing for 14 years. But it had halted it in 2003 under the concern, obviously, about an attack, about pressure and intervention. Whatever the intent of releasing this NIE in the last few weeks, the effect has been twofold. On the one hand, it has seemingly bought more time for diplomacy to work because it has said, 'You will face a dire choice on Iran in the future, but you do not necessarily face it this year.' So, it gives you more time to work with the international community to try to figure out if there is a diplomatic solution, if not a solution a way to at least slow down the potential of an Irani nuclear programme.

Second, it is also complicated diplomacy to put pressure on Iran to slow down. So, Russia and also, maybe, China to some degree can seize the NIE as evidence that you certainly do not need a third round of sanctions. Our European allies—especially the EU3 of Britain, France and Germany—are left somewhat dismayed at the lack of consultation over exactly what Washington is up to as it pursues this phase of non-proliferation diplomacy. Again, Sir Michael Quinlan will have more to say about this. But, however Iran's diplomacy comes out on this issue, it has profound applications again for the greater, wider Middle East and for global non-proliferation, just as negotiations in East Asia with North Korea where we have seen some progress but hardly the complete divulgence of the nuclear file that North Korea was promising to give. Do not expect that! Do not hold your breath on that! But at least there is a diplomatic track that has been pushing that ahead. That seems to show an alternative to the so far faltering diplomacy in dealing with Iran.

I will mention very briefly two other challenges. The fourth challenge is less of a direct security threat; it is much more of a broader set of questions that the strategic community seems to be debating. It really centres on what we could generally call the rise of Asia. This is certainly often seen in the Asia Pacific as a contest of a decline in American power and the rising China. Also embedded in it is the historical legacy of extant concerns, especially Sino-Japanese. But this can carry over into other regions, for example, the perceived competition between rising Asian powers, including India and China, over energy security and other issues around the world. I know Jack Gill is going to be talking about the complexity of energy security, later today, for the IISS.

Fifth, I think what has recently emerged, and this is clearly a mark of globalization, is that the global climate change has now become a mainstream security issue in the 21st century. It is threatening to create a new paradigm around which security is organized even with trans-national issues and global issues at the core. This appears to be a very slow train wreck in motion with nobody having a clue yet about how to stop it. Hence, there is a huge challenge of how to address the issue as well as all of these short-term issues regarding how to mitigate the damage that is here right now in our day and age concerning climatic change. These are challenges that I think will increasingly be on the agenda globally, not just in selected regions or countries.

Very briefly though, despite these daunting challenges, I am also struck by the tremendous opportunities that are out there, first for peace-making. Countering radicalization is only one of the obvious reasons for peace-making. But when we look at the Annapolis Conference that was held recently on the Israeli-Palestinian conflict, it seems to be too little and too late.

Clearly, there needs to be a concerted, persistent push on the part of the international community to try to find a just solution to the Palestinian-Israeli issue. I think Annapolis was a welcome addition to that. Even the Syrians broke out of their isolation. So, it shows the potential for engagement. Even while the peace is not going to be easy, the effort needs to be constant. I think it will be, at least from the US, in the coming years. Second, reduction of nuclear dangers, revival even of nuclear disarmament—Sir Michael Quinlan will speak about these things. I think that there are some great opportunities there for institution-building, for thinking about well beyond the Nuclear Non-Proliferation Treaty (NPT) regime, strengthening the various processes of trying to contain, stop, even push back, nuclear weapons.

Third, the rise of Asia may be even more an opportunity than it is a challenge in the same sense as in a book by my good friend, member of the IISS Council, Kishore Mahbubani, the subtitle of which is *The Irresistible Shift of Global Power to the East*, which will be published in February. He talks about three scenarios we really face and that we either have to help the East rise and make these tremendous contributions of taking people out of the clutches of poverty, of raising a much bigger middle-class, of contributing Asian creativity and inventiveness in industry to the problems that we face globally, not just here in the region, or we face a world in which we set up fortresses and protectionism and go the other way and have much more confrontation. That would be, of course, a loss for all of us. I think, as the Foreign Secretary said, the good news is that the big-power-level globalization has raised the stakes so much for the big powers to be confrontational. But what we have not seen yet are the common ground catalysts that push us all in the right direction to help make this happen. Nonetheless, this is an opportunity.

Grappling with climate change and other transnational issues I think, again, is an opportunity in the sense that we know that we built institutions that were very effective for security after World War II, but those institutions are not necessarily the right ones or effectively represented for the 21st century to grapple with especially these transnational and global issues. So, we have tremendous institution-building at both the global and the regional levels that arises because of globalization and the challenges I had mentioned.

Fifth, all of these, as I have just mentioned, bring the major powers into stronger partnerships in pursuit of common ends. At least they have the potential for bringing the major powers into stronger partnership, not least the US-Indian strategic partnership.

Let me just add one closing idea about US as one of the major powers and as a dominant power. It is still the dominant power but it has lost its influence. Because of that, I think the major challenge of the next administration in Washington will be—whether it is Democrat or Republican—to start a foreign and security policy that is predicated on how to regain US legitimacy or at least a modicum of legitimacy, how to regain some of the intellectual leadership on this agenda. That is, I think, good news because it is a recognition that the US cannot do it alone and, therefore, I think you will see US policy that is reflected by greater strategic restraint, accepting deterrents and imperfect security rather than seeking issues like preemption or quick resort to military power. I think you will also see much more concerted diplomacy, what Democrat and Republican alike are talking about as the 'smart power', the weaving of soft power and hard power more effectively together, that is, using soft power.

But it will require a more intellectual leadership and a willingness to work in concert with other powers. That is where it will depend on the leader himself. We will have to see that in 2009, when the US takes that direction. In the meantime, I see the Indians think that you will be a world power. You will be the world power in 2020 according to a front page Indian survey today. So, do not wait until 2020 to start contributing even more to this debate. The time is now. India has already really risen. I think it is a welcome ally to not only grapple with the challenges that I mentioned but also to seize the opportunities as we really need some leadership.

Thank you very much.

Chairman: Thank you Patrick. May I now invite Dr Ummu Salma Bava, who is Professor for European Studies and coordinator of the Netherlands Prime Minister's grant at the Centre for European Studies, Jawaharlal Nehru University, to make her presentation?

Ummu Salma Bava: Thank you Mr Dasgupta. I would like to thank you and your division in particular for inviting me to take part in this Foreign Policy Dialogue. Let me say at the outset, I speak here as an academic and I do not represent the Government of India, and the official view has already been presented this morning. Patrick has already flagged the challenges. India is arriving on 2020. Is the show over? Should we have another coffee break? But I have to follow my brief and my brief was to present a larger strategic canvass and not address the particular issues that are going to be taken up in the next two sessions, and that is what I intend to do.

Let me flag at the outset that a thing that we are witnessing today is that patterns of order and familiarity of structures have given way to uncertainty, challenges and, more importantly, opportunities for all states. I would like to address this as an interplay of a four-set matrix, which I would identify as follows: (a) structures, (b) actors, (c) institutions and (d) issues. Patrick has just underlined it. I think what we knew as bipolarity has shifted to a period of unipolarity or a uni-moment. And what we see currently is the emergence of some kind of a multi-polarity. I say that because we see a diffusion of power at the military, political and economic level. It is this really three-dimensional chess set which makes it very interesting as to how the structure of power will get arranged.

If the balance of power that we were comfortable with has given way to a preponderant military power held by one, we are also witnessing new economic power centres that are emerging. The other shift that is identifiable is the move from a defined identifiable threat to a diffused and a diversified threat. All of this structure, which I have just identified for you, is being impacted by the twin forces of hegemony—which I would say underlines concentration—and globalization, which I would say is equivalent with the fragmentation that we are witnessing.

On the level of the actors, I think what we see interestingly is the juxtaposition of the old and the new. The old actors are there. When we look at the global governance system, we see the presence of the old. What is significant and interesting for our discussion here today is the rise of the new actors and I would like to flag the economic rise of India, China and, say, the Asian tigers. But more significantly, along with these economic heavy weights is also the political rise of what I will call 'Brand China' and 'Brand India'. I think that is something that needs to be brought out.

When you start looking around the globe and look at these new kind of structures and groupings which are coming together, you really do get a veritable alphabet soup. You have got the BRICS (Brazil, Russia, India, China, South Africa) on the one hand; you have got BRICSAM (BRICS, ASEAN, Mexico); you have got emerging powers, markets, pivotal states, central states, and what have you. We do not just stop at the states over here. What we also increasingly get is the non-state actors, corporate groups, transnationals, civil society, and you have got a whole network of criminal and terrorist groups which are working in collusion with each other.

When you come down to the institutions, we have got the recognized international institutions; we have got the UN, World Trade Organization (WTO) and the international regimes. But significantly, at the issue level, all of them are involved in, what I will call, global governance. Whether it is global governance regarding transnational security, whether it is multilateral trade arrangements, or conflict resolution through the UN system, each of them in some way is about putting an order or a structure in place so that you can have predictable outcomes of behaviour when states interact with each other.

The question which is being asked today in this shifting landscape is will the nature of global strategic framework lead to more and set predictable outcomes in the future? I think that is a critical question that is coming up. As a consequence of this interplay of the four-set matrix of structures, actors, institutions and issues, the outcomes I would like to flag over here are four. First, there is a changing strategic calculus which is especially being driven by Asia. Second, Asian economic dynamism is providing global stability but some of the major geopolitical conflicts are also generated from this region. Third, there is a new geopolitics and a geo-economic gravity which is emerging. From an economic, political and security perspective, I would call this the rise of the Chindia factor, that is, the China and the India factor, in geopolitics today.

Having said that, I would like to present to you the challenges and opportunities for this four-set matrix, that is, structures, actors, institutions and issues. I would submit to you that we should examine this within two global contexts, which are hegemony and globalization. These two factors were also flagged by the Foreign Secretary this morning. What we have juxtaposed is on the one side, a concentration of military power and on the other side, when we talk of globalization, we get networked interdependence. If from the military point of view we talk of a balance of power, then economically, we are talking of the power of dependence. Politically, as a consequence of this, you are getting a mixed result depending on the mix of the hard and soft powers that states have and can use.

What is the result of all this impact on the structure? First, the structure is in a flux. Do we know a definitive outcome? Is there a blueprint? Is there an architecture? I think there are big question marks over there. When you look at the actors, I think the old and the new are either in competition or they are in a cooperation mode. I think it is critical that what you are getting is a cooperation on interest or that issue-based cooperation is taking place. You are not going to get the aligned structures of the old that we have known. When we come to the institutions, I think they are stretched to capacity and some are unable to deliver any more. The question is, do we have yesterday's institutions to address today's problems? Do we have the requisite capacity for them to lead through in this? Each issue that Patrick has also just highlighted is increasingly becoming transnational and securitized.

This really brings the question over here—if you are talking of the strategic shape of the world—as to what kind of a normative template of order and governance is globally going to emerge. I think one can say that there is enough compelling evidence in argument here that there is a diffusion of norms which is being challenged both by existing political systems and emerging powers. I do not think norm-recipiency is happening in the same way as it happened right after these norms were put into place in 1945. I think that is being challenged. I think there is also, increasingly, a contestation of ideas, of politically organizing principles. Should the one norm mode be there? Is this the only way we are going to organize collectively? Is that only way being endorsed? I think it is going to be that. I think a critical factor from the point of view of emerging powers is the issue of who creates these benchmarks. I think that is going to be a critical debate in the time to come.

When we talk of benchmarks, there is the issue and the problem of a security trap and a new agenda, as I see it. I think what was significant from 1945 to 1990 was that a state or a group of states were either security providers or balancers or guarantors. Today, what we see here is this role of security dispensation which is being challenged, challenged in terms of the ability in the North, that is, say, within the European Union (EU) can the EU be a security provider, balancer or a guarantor minus the NATO? If you come down to the global South, to the new emerging actors who would like to be in a larger global position, do they have the capability and the capacity? Is that the kind of role that they would want to take on? That is still kind of open.

What we definitely see and will witness in the future will be a greater contestation for power and leadership coming from different parts of the world. It is not going to be the North-led global order. In other words, there is a shifting hierarchy of power between states. I think what we would increasingly have to contend with, either as a community here or as policy-makers, would be the shift from the notion of waging war to waging peace. Whose normative notion are we going to endorse when we say that we are going to wage peace now, which is very different from peace-keeping and peace-making? This would also involve elements of the pre-emptive, of going beyond the current norms of international law. Will one-rule-fit-all apply? Are we going to be singularly deciding on a case-by-case basis? That is something which still has to be discussed and discussed in greater detail because these would be the new benchmarks that would come into place. Is the world ready for it at this point? I think we have seen India's reticence on this. It is clear that India is not going to jump in just because somebody said that this is the new thing we have to do or it is the current flavour of the season.

This brings me to the current debate we are increasingly seeing about creating, what we call especially in the context of the emerging powers, responsible stakeholders and the whole notion of global burden sharing because this is kind of linked to democracy as a norm of the compliance benchmark, as I would like to call it—democracy being put out as the political organizing tool and that if you have to be the good actor in international politics, then your internal organizing principle should be democracy. Just shifted to the global level, you hardly get a democratic system in terms of global governance. That is clear. Even as we speak, when we look at the outcome of the Bali Conference, you can see how in this kind of partnership of sharing, what will be the impact on climate by the different stakeholders is differently applied. The US is not willing to have any kind of a number put down to, say, emission standards. I think that is there.

Having said that, let me just come down to the last two points I would like to present to you and that is to ask what is the template for the emerging powers for countries like India and China? I would say that the template is still the dominant power. The Foreign Secretary this morning said that the Westphalian system of the state is kind of dying or disappearing. I do not think so. I think there is a strong emphasis on retaining that at least in the global South. They are very modern in a Bob Cooper's kind of definition. They are definitely not post-modern in the EU sense of the word. I think the two questions which are posed here are as follows: (*a*) how will emerging powers like India behave in international politics and (*b*) will the rise of the Asian powers be different from the Western powers? It is more important to ask, should it be different if your template is the US?

Let me conclude by making four quick points on India as the new emerging actor or the emerging power over here. Patrick said that by 2020 we shall be there and he has just advised us to move on, push the gas and get on there faster. I think from a domestic context the real challenge for India is that it rides these two horses of being the developing and the emerging power simultaneously. It is the leader of the South going North. As the leader of the South going North it is confronted by three questions which probably the global South would like to pose to India. Will it speak for the South and continue to speak for the South? Will it be an inclusive actor? How do other states respond? I think to reverse the question and ask from a North or a Western perspective would be to ask, is India emerging as a recent norm setter in international politics? Will India be, as the West would like to ask, a responsible stakeholder in global burden-sharing? When these two questions are posed to India, what is being held out is an action list of compliance that if you comply with A, B and C, then this is the litmus test for India's credibility.

I would like to just conclude by saying that democracy is not new for India. When we talk of the democracy *mantra* and the organizing principle, India had opted for democracy with a population which was still not literate, endorsing a principle to bring about change and conflict resolution at the domestic level from 1947. We have seen a smooth transition of power from one government to another which talks about how India has believed in the value of democracy. It is not something new that we will now stand up and endorse. From an Indian point of view, what India would look at in terms of global governance is to look at an equity-based global governance structure. That is extremely critical. When you look at global governance structures today, equity is something which is really lacking over there. From the perspective of the old actors, the question is, are they ready to adjust to the rise of the new actors?

Thank you.

Chairman: Thank you Dr Bava. Both the lead speakers were obliged to take a canopic overview and make a canopic presentation. We start coming down to somewhat micro issues now. It gives me great pleasure to invite Sir Michael Quinlan as the first discussant. He is the Consulting Senior Fellow at the IISS, London; he spent several years in the British civil service and his final post was that of Permanent Under Secretary of State in the British Ministry of Defence. The area of specialization for him has been essentially on nuclear issues, in particular nuclear weapons. Sir Michael, you have the floor.

Sir Michael Quinlan: Thank you very much. It is a great pleasure for me to be here again in Delhi in dialogue with one of the liveliest and most gifted seat of communities to be found, I believe, anywhere in the world. I should make it clear that I no longer speak in anyway for the British government. Her Majesty has been managing without me for over 15 years now. I intend to focus upon the nuclear weapons scene partly because that is my subject but also because, healthily, it is less salient than it used to be; it could become disagreeably salient again if we get it wrong.

I would just not intend to stay around the non-proliferation regime. I am very well aware that India is not formally part of the NPT regime but nevertheless, I am quite certain you would agree of an Indian interest in the subject.

I do not myself take the apocalyptic view which some do that the regime is in very grave trouble and on the verge of collapse. I think it remains a pretty strong structure and its continuance is plainly in the interest of the vast majority of the international community. But there are both general weaknesses and particular aspects to it at the moment.

To take the general weaknesses first, I would identify three. One of them is what are called the safeguards which constrain what states do, are not everywhere comprehensive and in addition, the verification systems for ensuring that the safeguards are observed are also imperfect as we found very uncomfortably when we discovered in 1991 what Iraq had been doing. There is an additional protocol, which the International Atomic Energy Agency (IAEA) has devised but that is by no means yet universally adopted. So, that is weakness number one.

The second weakness is the problem of withdrawal. It is possible, as you know, to withdraw from the treaty just on giving six months notice and some story about why. It is not possible or practical to amend the treaty. But I do think the international community needs to construct a better system of penalties, deterrent penalties preferably rather than post-event punitive penalties to make withdrawal a bigger and more difficult matter than it has been, for example, for North Korea.

The third weakness is what I might call the threshold problem by which I mean that states can without necessarily breaching the treaty place themselves, whether deliberately or incidentally, in a situation in which they could move really quite quickly to the development and acquisition of a nuclear armoury. That is the third problem. I think it is going to become more and more important as the likely spread of nuclear energy for a variety of reasons takes hold of more and more across the world. These are the three weaknesses which the treaty review conference will have to confront in 2010, and the regime certainly cannot afford to have another fiasco (and I think that is not too strong a word) in 2010, like the one that happened in 2005. I do not suppose that 2010 is going to resolve all the three weaknesses comprehensively. They are all complex and they are political. But I think it must make some headway, if the regime is to maintain its credibility.

If that is to be achieved, a particular responsibility obviously falls upon the five formally treaty-recognized nuclear powers. That brings me to the question of Article 6 of the Treaty, which, as you will recall, commits the nuclear powers to move more towards eventual disarmament. It is quite often suggested that that is the only bargain in the Treaty. In my view, the Treaty has several bargains. This is one of the them, even if not the most specific, and it has to be taken seriously if the five are to have the political credit and the moral authority to make headway with correcting the weaknesses which I have referred to.

Most of the five—China is the exception for reasons which one can understand—have done quite a lot in the direction of disarmament, much more, I think, than is commonly recognized. The US has made massive reductions in its armoury. It is not developing new weapons for new purposes. It has removed a whole host of categories of nuclear weapons. Perhaps, it will be well to make all that slightly more effectively understood than it has been. Britain and France in their much smaller base, of course, have done a great deal in the direction of disarmament. My own country has recently announced a further reduction in its operation, holding warheads even below the relatively very small level which it was actually for.

But I am sure that more could be done, especially by the two biggest armouries. I would hope to see the current Moscow Treaty, which is a pretty odd and loose kind of an arms control treaty, replaced in due time and in a timely manner by something more rigorous and more extensive, perhaps, for example, bringing under some kind of control the very large numbers of non-strategic and nuclear weapons which Russia still holds, which are under no kind of control and are supposed to be many times larger, because Russia makes very little information available, than the holding which the US now has in that category.

Behind all these there is still the long-term aspiration to move to a world in which there aren't any nuclear armouries. That is something in which the Institute has been taking a special interest. There is a commitment implied, both in political rhetoric and in, for example, the preamble to the NPT. But there has been, I think, surprisingly little serious examination of what would have to happen for that to become something other than just a rhetorical phrase. The Institute has embarked upon a study, not with any policy intention, just to elucidate, to test what would have to happen. We should be bringing that to this dialogue. When Mr Arif Khan brought his team to the UK earlier this year, we talked about this subject and India showed a clear interest. You remember that Rajiv Gandhi himself in 1988 reminded the world of this goal. That interest has been a real stimulus to our carrying this exercise forward.

Let me talk just briefly then about the particular threats. There are two of them—we are familiar with them—the Democratic People's Republic of Korea (DPRK) position and the Iranian situation. I want to talk primarily about Iran because I regard that, for a variety of reasons, much more serious of the two. The recent US intelligence estimate clearly takes the heat off the subject at least in terms of time. From one standpoint, I welcome that, in that I think it removed any likelihood of drastic military action by the US, an action which I would have thought as close to madness as anything I can recall with the possible exception of the invasion of Iraq. But, the intelligence estimate is not a certificate of innocence, not by a very long margin, or a certificate that worry is now over. Iran had a serious clandestine programme of weapons development in gross, flagrant breach of the treaty. There had been a huge amount of concealment and men-dacity, and that continues. Iran is ignoring United Nations Security

Council Resolutions. It is very hard to interpret its determination to maintain its own national enrichment capability and to reject all offers, and there have been several credible ones, of international help with that. It is hard to interpret that it has been anything other than a determination at least to proceed to what I have called earlier a threshold capability.

That is bound to be worrying. Iran is a very big and important power in the region, far more than DPRK is in its own region. It is not a status quo power. It is not a force for stability at least at present. It is against the existence of Israel and we have had Mr Ahmadinejad making a remarkable, and to my mind gravely objectionable, remark about that. It is clearly supporting, well if I will not call them terrorists, let us say violent non-state actors in the form of Hamas, Hezbollah, and the like. If Iran moves even to an evident threshold capability, that will stimulate others in the region to think that they must take their own insurance perhaps in the same form, and more generally, for Iran to escape or semi-escape from constraints of the regime will do serious long-term damage if such an escape is made cost-free.

They may in the end have to acquiesce. There is no guaranteed way of preventing Iran from becoming either a threshold or possibly even an actual nuclear weapon power. But that will be a very bad thing. I think the international community needs to do all it can to prevent that from happening. That means that we must maintain pressure as well as, no doubt, construct some rather better carrots and sticks than have yet been made clear. I think we have to consider what we would do if Iran does go all the way but I do not believe we should in any way give up the attempt to prevent that, and that does mean, I am afraid, that we have to maintain pressure.

Thank you all for being patient.

Chairman: Thank you very much, Sir Michael. Could I now invite Dr Manpreet Sethi to make her presentation as the last discussant for this particular session? She is a Senior Research Fellow with the Centre for Air Power Studies and heads the Nuclear Security Division of the Centre.

Manpreet Sethi: Thank you so much, Mr Dasgupta; it really is a pleasure and a privilege to be participating in this conference. I think I must confess that we have grown up as a student and other-wise, reading Sir Michael's papers. So, I am really delighted to be speaking after him though I do not think I would be contributing anything much beyond what he has already said.

What I am trying to look at is the strategic shape of the world through the nuclear prism. That is the brief that was given to me, to concentrate on the nuclear aspect of how the shape of the world looks like. I think by the end of it, you will be able to take your pick on whether the glass is half full or half empty depending on what you are putting into the glass, whether it is proliferation or non-proliferation.

I want to start by looking at some of the major global nuclear issues which confront us today, which aren't very different from when we started in 1945, as was mentioned by Dr Patrick here. Nuclear deterrents, nuclear proliferation, energy and disarmament, and their inter-linkages are boiling the nuclear cauldron all over. As I go along, you would be able to see these inter-linkages a little more clearly—how deterrence is feeding into proliferation, proliferation is also getting fed in through nuclear energy, and nuclear disarmament, being so much not-fashionable these days to talk about, is leading to the kind of problems that we are facing in terms of the strategic shape of the world.

If you look at the state of nuclear proliferation, the traditional concerns since 1945 were primarily horizontal and vertical proliferation—horizontal in terms of the increasing number of states which were going nuclear and vertical as in the increase in the number of nuclear weapons. The traditional responses to these challenges were essentially treaty-based arrangements, whether it was NPT, Comprehensive Test Ban Treaty (CTBT), or the other kinds of treaties that were put in place; export controls or tech-nology denials to keep the countries away from nuclear weapons technology and universal compliance of the NPT was largely the *mantra*, which was put out as the panacea to proliferation. If every country becomes a non-nuclear weapon-state member of the NPT, it would end proliferation.

But if you look at the state today, the membership status of the NPT is 188 and only three countries are out. I would even go to the extent of saying that those three countries which are out are today the minor problems that the NPT has to deal with. There are several other gaps that we will be talking about. The universal norm of non-proliferation is not existent. There is the threat of proliferation, which is acute. Proliferation is a stark reality despite treaties, despite sanctions, despite supplier controls and despite the threat of military strikes.

I think even in the case of Iran—if I could differ a little bit with Sir Michael to suggest that these are not going to be the ways in which you are going to be confronting the problem of Iran and resolving it in any big way—we will have to look for situations and opportunities beyond just the treaties, or the sanctions or the threat of strikes. These have not worked. We will have to look for solutions outside.

Why are we in this state? Why is the NPT in the state that it is today? It is largely for two reasons: (a) it is the result of the selective proliferation by the treaty adherents themselves. In this, one would hold the nuclear weapon as well as the non-weapon states of the NPT equally guilty; and (b) for commercial gains, as has been demonstrated with the A.Q. Khan network. Very often we just tend to look at the A.Q. Khan network as the A.Q. Khan and the non-state actors. It goes much beyond that. All the countries which were non-nuclear weapon states within the NPT or nuclear-weapon states having export controls in place were responsible for the kind of network that A.Q. Khan managed to engineer. Also, for strategic gains, we have seen selective proliferation happening. I think this is exemplified in our neighbourhood with the kind of proliferation—nuclear and missile—which has taken place from China to Pakistan, going to the extent of handing over nuclear-weapon designs to Pakistan.

The other reason why the NPT is in the state it is, is because of the result of the rising salience of nuclear weapons in the strategies of nuclear-weapon states. So, what you have are in the doctrines and policies of all the nuclear-weapon states—there is an emphasis on enlarging, on modernizing, on creating new thinking on nuclear weapons, making them more useful with the kinds of threats that exist today. One would particularly like to mention UK because so recently in the past it was one of the countries that was being looked upon as leading the way on abolition in terms of not wanting a deterrence for any security reason. But it has reinforced its faith in deterrence and it is going ahead with modernization of the arsenal. The next reason is the demise of the hope for disarmament. The larger cynicism has afflicted the debate on disarmament; I think all these four reasons contribute to the situation that the NPT finds itself in today.

The situation as it exists on proliferation is that there is vertical proliferation in terms of what we hear about the Reliable Replacement Warhead (RRW) in the US and the Ballistic Missile Defense (BMD) and how that is triggering off a chain reaction—horizontal proliferation—Iraq, DPRK and Iran, the threat of Weapons of Mass Destruction (WMD) terrorism, again the use of possibly fissile material by non-state actors to indulge in some kind of a terrorism, which becomes more possible with the kind of globalization that we have been talking about since morning, and the more diffused and diverse sources of proliferation which exist and which A.Q. Khan was able to use. So, the contemporary nuclear reality is that there are gaps in the existing non-proliferation architecture. The NPT does not have any mechanisms to deal with non-state actors. That is something we will have to be looking out for.

Article 4, as again Sir Michael said, refers to the problem of the threshold states. Under Article 4, the country can reach the level which would make it very easy for it to transcend to a nuclear weapons capability. In Article 6 as well, I think there is a gap because there is no commitment which is visible from the side of the nuclear-weapon states which pushes the others into all kinds of insecurities and creates the motivations for nuclear weapons.

There is a need for immediate remedial measures. But what is happening is that there is no patience with consensus-building. Nobody wants to look at the NPT in terms of changing it in any way because it is seen as a big exercise. There are no other alternative arrangements which are being talked about. Instead what you have are coalitions or willing kind of arrangements. So, where country interests coalesce in a certain way the coalition or willing arrangements work. We have seen this in the form of the Container Security Initiative (CSI), the Proliferation Security Initiative (PSI), the Thousand-Ship-Navy, and the UNSCR 1540 where the five countries got together to put out this security resolution and imposed it on all the countries to have mandated that acquisition and possession of any kind of WMD material would be unauthorized.

There is a global initiative to combat nuclear terrorism. Again, a small group of countries have got together. Then there are proposals such as the Global Nuclear Energy Programme (GNEP) from the

US, the Global Nuclear Power Infrastructure (GNPI) from Russia and the Fuel Bank from the IAEA Director General which is meant to constrain countries from having full mastery over the fuel cycle.

Another new measure of non-proliferation that I would like to flag is the Indo-US Civilian Nuclear Cooperation, which again is in a sense trying to look at the problem of a state which is outside the NPT but which upholds the cause of non-proliferation and how that can be accommodated into the regime.

India, I do not need to even say, has been a consistent supporter of the cause of non-proliferation despite it being a critic of the NPT and technology denials, which we have found are ineffective unless they are holistically implemented with uniform rigour and uniform sort of sincerity and used without political malfeasance, so there is no political motive or manipulation of the regimes which takes place. India is keen to participate in the regime to the extent that it can but it is constrained by the nature of the NPT because you can neither be a nuclear-weapon nor a non-nuclear weapon state under the present definition in the NPT. So, the alternate route of participation I think came out in the form of Indo-US Civilian Nuclear Cooperation Agreement, which would bring in India in the larger non-proliferation picture. India's accommodation has largely been on the basis of its proliferation record; the energy needs of the country are being understood and we have seen how the environmental concerns have been so properly voiced in the Bali Summit to the extent that one understands why the kind of energy needs that India has need to be met with the help of nuclear powers as one component of the larger energy mix.

Indigenous nuclear technology expertise is also beginning to be recognized—that India has followed a certain path on nuclear technology development, which is different from the others. Reprocessing, fast-breeder reactors, use of thorium, are now beginning to be seen by the outside world as alternative routes and which make India and its programme more acceptable.

Then, there are commercial benefits, of course, in terms of the market that India offers, whether nuclear or in other high-end technology goods, as well as strategic interests which are not spoken about so much in India as outside, about how this kind of an arrangement would help to contain China in a certain sense. But I wanted to flag these as the possible reasons and why India is being accommodated into the non-proliferation regime now.

Coming back to the future of non-proliferation in the larger perspective, I think till the time nuclear weapons are around, what we will have to do is to manage proliferation as best as we can through laying down of norms through treaties. To the extent that countries want to accept them, they will. But there should be norms which would be laid down. Creation of supplier controls, enhancement of the safeguards regime to the extent it is possible, interdictions to halt illegal commerce so that activities like the PSI or the CSI would be in place, better security of nuclear material and an international control over the nuclear fuel cycle—these are all the mechanisms that will be used in the future to handle proliferation as best as we can.

If we want to also look at promotion of proliferation resistant reactors—again, research and development, which is going on—if we want to look at an alternative scenario to get out of this vicious cycle of deterrence, proliferation, etc., if we want to ensure sustainability of non-proliferation, it will need to be done through a much larger set of activities. One is to improve international security, [achieve] reduction in the role and salience of nuclear weapons. Can we in some way reduce the importance of nuclear weapons in national security strategy so that they do not seem temptingly attractive to other countries? [Second], de-legitimization of nuclear weapons—I think India has been at the forefront of this debate about whether we can de-legitimize nuclear weapons in any way, and we had a window of opportunity in the early 1990s where there was a huge amount of optimism about how nuclear weapons will fall into disuse and become the detritus of the Cold War. [Third], acceptance of the no first use doctrine—if this gets accepted in a way by the nuclear-weapon states, it automatically leads to the nuclear weapons falling into disuse.

All of this, of course, requires change in mindsets and belief systems. Nothing can be done unless the idea takes root in our minds that this is the way it should be looked at. It might take a long while to get there but it is worthwhile to take that decision to move in that direction. So, a nuclear-weapons-free world, I think, is critical for sustainable non-proliferation.

Thank you so much.

Chairman: Ladies and gentlemen, we will now open the floor for discussions. What we hope is that the inaugural address by the

Foreign Secretary and the four speakers who have spoken just now would trigger off a timely discussion. My only request is that if you are asking a question or would like to say something, please introduce yourself. Thank you very much.

R.B. Mardhekar: I just wish to make two small points. When you talk about a global response to a problem, whether it is nuclear proliferation or anything else, a cohesive global response will be possible only when nation states are themselves cohesive. What we are finding is that increasingly, economic models and political systems are leading [and] rather than generating cohesiveness among nations, are creating divides and segmentation within nations. In a situation like that you cannot have a global cohesive response to any problem.

Second, I personally believe that when you talk about the shape of the world to come, this session should have had a separate session on the impact of religion on strategic thought. We talk about terrorism and you find that most people who talk about terrorism are focusing on Islamic terrorism. The phrase often used is 'distortion of Islam'. I do not think you can ever tell a believer that he is distorting his own religion. The moment you start doing that, you are generating grievance and anger. The more you focus on a particular religion when you talk about terrorism or global issues, the message you are sending out is that religion or those whom you claim distort the religion have managed to achieve what they wanted to achieve.

In the old days when terrorism was talked about in political terms, whether it was the Red Brigades or Baader-Meinhof, it was easier to handle. It is like telling a Catholic that the Pope should not live in a palace. Or it is like telling a Hindu that there should be no temples—that temples are contrary to religion. You cannot tell a believer that he is distorting his own religion. The moment you start talking along these lines you are in fact generating anger. The second thing is isolating one religion and trying to come to terms with it. It does not work. We have seen the US trying to have a dialogue with Muslims. The Muslims are not the problem: terrorism is the problem. It is not necessarily confined only to Muslims. You can have Hindu terrorists; you can have Christian terrorists; you can have any number of terrorists. The issue has to be dealt with as a political issue even though it is shaped by religion.

Take, for example, Iran. We talk about proliferation in Iran. I had the opportunity to serve in Iran for a long spell during the Revolution. The Revolution was seen as an Islamic movement and a religious movement. I beg to differ with that view. It was entirely a political movement couched in the idiom of religion because that was the way to gather strength. Iran's policy, whether it was with regard to the Hezbollah, whether it is with regard to nuclear non-proliferation, to my mind has been a defensive policy born of a historical memory. They are surrounded by states they considered enemies. Sunni states, Wahabi states, have always considered themselves under threat traditionally. A lot of the actions of the day, including the rhetoric they indulge in and the exhibitions they indulge in, is born of the fact that they believe that the world has ganged up against them.

At the time of the Revolution, the same thing happened. You have the whole world saying this was bad, whereas the people on the streets were saying this was the right thing to do. You are seeing that in Pakistan today. You have General Musharraf, who is supported by the entire world because he is supposed to be fighting terror; yet, the people of Pakistan on the street are saying, 'We do not want him.' The question is, what does the global community listen to? If you want a cohesive answer to the problem like proliferation, the problem like terrorism, you cannot adopt policies which segment nations within themselves.

Thank you.

Arundhati Ghose: I am sorry I missed the Foreign Secretary's speech but I was very interested in hearing the discussion today. I have one question. I am not quite sure who is likely to answer that, maybe Dr Cronin. I did not hear any reference to Russia. When you are talking about the strategic shape, Russia is today probably the most puzzling. We do not quite know where it is going, how it is going? People in the West seem to be worried about the rise of China. They are not worried about the emergence of India, though. Whether it is emerging or not, 2010 or whatever, that is another issue. I do not think we need to bring that into the same league. I think people need to include the issue of the transformation of Russia in the next 10–20 years, which will determine its geographic shape. So, is there any reason why Russia is not referred to? We have talked about, Professor Bava also mentioned it, the East, the rise of the

China–India aspect. This is all overdone. China, yes. India is not at a stage where it threatens anybody. Russia, we need to look at what is happening there. That is the question. I am not quite clear as to why it is not taken into account.

On the nuclear issue, a large number of issues have been raised—I just finished [reading] last night a book called *Deception: Pakistan, the the United States and the Global Nuclear Weapons Conspiracy* [by Adrian Levy and Catherine Scott-Clark, Penguin India, 2007] about the way in which Pakistan's proliferation programme was slated to go forward by specific actions of the US. Of course, this is written from the point of view of the US but everybody was involved—I think that we need to have perhaps a separate session on the nuclear issue. That is because WMD and terrorism, which, certainly as an Indian, I would be extremely worried about—I am very worried about it. We would like to hear what exactly is happening rather than go back to the NPT and what you should do and what you should not do. We have got a problem right now. This is a comment and a question just in case Sir Michael would feel it worthy to answer.

Thank you.

Patrick Cronin: Well, these are very good comments and they point to the difficulty we are in, which is trying to put too much into a single short panel and we apologize for that. I certainly also did not derive the name of my presentation I was assigned. In fact, it somewhat surprised me—*The Strategic Shape of the World*. Of course, Russia was implicit in the issues that I was hitting upon—the themes from proliferation to talking about the big powers themselves. I think it is a very important issue. It was what I put very much on the front and the centre of the IISS's research agenda and it has been recently published in a Delphi Paper by Eugene Rumer. In many ways, Russia has been a big power for 300 years. They look as an aberration in the fall of the Soviet Union and the complete oversight and marginalization of Russian power and influence. Now, they are fighting back to say, 'Wait! No, we are a big power and we will act like a big power and will use primarily our energy card as a way to exercise this power and influence.' Certainly in the EU–Russian energy relationship, you see it enacted on the Europeans with a vengeance in terms of pursuing pipeline politics over the Eastern Europeans, directly to the individual European nations, especially in the West. But you see it in many other ways, the reconstitution of the Mediterranean fleet deployments by Russia and just the rhetoric, talking about the nuclear parity the other day. So, Russia is definitely a real player on the scene and that was my shorthand of big powers—it was included within that.

Just briefly on the issue of religion, this is always a tricky question. Of course, to discuss religion and to discuss the relationship and the inter-linkages between religion and terrorism or extremism would need a much longer session. Of course, Nigel Inkster will treat with much better subtlety and nuance some of the issues I highlighted. I was quoting the Saudis when I talked about distorting Islam. So, it was not a Christian talking about Muslims. I can quote lots of sources on this. I was also quoting Edna Fernandes, who has written the book about how we really should not associate religion with terrorism, quoting from different sources, than to highlight the fact that in particular, people in the strategic studies world are focused on a particular set of problems, namely, the Al Qaeda network, which has had its roots in Islamic base, no pun intended. That was simply what I was trying to highlight. Again, I will defer to Nigel, who will, I am sure, be able to add to this or undo some of what I have said here later today. Michael, do you want to talk about it?

Ummu Salma Bava: Just in response to what Ambassador Ghose said on Russia, I flagged it as actors old and new. Probably I should have brought out that there is a new Russia, a resurgent Russia, driven in the sense of this oil pricing, which gives it leverage. I think what will be interesting then to see is how it uses this leverage in the realm of ideas to create institutional structures where it is asking for a new kind of a parity based on an old position of the great power, which it has been. In that sense, what would be interesting is would we see a new kind of a bandwagoning taking place where Russia would come in with the Southern actors? Or would it see itself primarily as still this Western, Northern actor, but going through a transition period where it has to rewrite its internal agenda a little bit and put in things? Putin is doing that with a new president coming into place soon and him coming as a prime minister. I think the interesting contestation would be in terms of the ideas area. Would we see the challenge of ideas from the South co-opting with Russian support or will it be that they are going to be seeing themselves as part of the other bloc? I think that would be significant.

Chairman: If I could just draw Sir Michael and Manpreet's attention once again to what Ambassador Arundhati Ghose said. The book *Deception: Pakistan, the United States and the Global Nuclear Weapons Conspiracy*, published by Penguin India, is doing the rounds these days as it has raised some critical issues that are a cause for genuine concern. I think one of the most critical points it tries to make is that the nuclear programme in Pakistan was consciously supported by successive US establishments. It was not as clandestine a programme as people might make it out to be. I think this is particularly worrisome. The deception is not so much whether there is a huge degree of shame and horror at what happened in Pakistan, but at the fact that it was allowed to happen and even actively supported by Western governments. That really is the issue. I was wondering whether Manpreet would like to take this up.

Manpreet Sethi: On the A.Q. Khan issue, I agree with Sir Michael that we need to look ahead and go beyond this. But I think if we do not look at what happened in the past, we might be surprised again in the future. Despite the fact that Pakistan claims it has a certain amount of security measures now in place, that certain amount of greater transparency has been built into the systems, that a thing like this could not happen in the future, there are, equally, other reports coming from the US officialdom, which also claim that the entire network has not died down, that other people have taken the place of A.Q. Khan, that he was one lynchpin. Given the fact that there is no investigation which is being allowed by the Pakistani administration and being agreed to by the others who were players in the field, one never knows what all has already gone out by now, to what destinations it has travelled, and how it would surface sometime in the future. So, while it is time to look ahead and look at the further challenges which are happening, I think it is equally important to understand and get to the root of the problem here as well so that we are not surprised again sometime in the future.

Since I do have the mike now, I wonder if I can just ask a couple of questions. Sir Michael, you said that there is a possibility or you see that there could be a replacement of sorts, of the Moscow Treaty that you mentioned, at some point in the future. Given the understanding that we all have of Russia's assertiveness politically, economically (now), its increasing threat perceptions vis-à-vis the BMD and the way it is reaching the shores of Russia, the Munich speech of Putin, all these are indications to some extent of the assertiveness of Russia, which is very well recognized. In that context, at a time when it is moving away from treaties—it has gone out of the treaty on Conventional Forces in Europe (CFE); Intermediate-Range Nuclear Forces (INF), again, it says, could be suspended at some point in the future—are they going to be looking at a replacement of sorts? Would it suit their interest to come up with any such thing?

The other thing that you had said which triggered off a thought in my mind was about the NIE estimate taking off the heat in terms of time on Iran. I think the heat was generated more by the time constraint of the Bush administration rather than a perception that Iran was immediately at the brink of a nuclear weapon. So, it was Bush's time running out, which was actually where the heat came from. I would like to have your perspective on that as well.

Ishrat Aziz: My own remarks regarding this issue are the result of what Amit Dasgupta said about the book *Deception: Pakistan, the US and the Global Nuclear Weapons Conspiracy.* I have not read it but I must read it after I have heard of this book.

They say: 'It is good to be smart but it is also smart to be good.' I change it slightly and say, 'It is good to be smart but it is better to be good.' I will say this because again and again I suddenly find that you see something being said like this programme was there, it was known, but now suddenly it is being played up. I think this approach of, 'He is a bastard; but he is *our* bastard'—that should go. What Franklin D. Roosevelt said about dictator Samoza—I think this happens again and again. I am saying it because we must realize that people understand these things. They realize that suddenly certain things are said and then suddenly certain things are suppressed.

As regards Iran's nuclear programme, we heard one thing. Now, suddenly their intelligence estimates say something else. But people still say, 'No, no, actually this intelligence estimate is not good; maybe it is politically motivated.' My basic point is that let us be fair and honest in these discussions.

Again, the same thing about the use of religion—I have no doubts in my mind. This problem of terrorism, of religion, of extremism and religion is something that has always been in my mind. I have had to think deep and hard about it. I have no doubts in my mind that religion is used to hijack frustrations born out of injustice and humiliation. It is hijacked, it is not the cause. If you look at terrorism globally, you will find it is there in different religions and in different circumstances. I have been hearing the discussion with great care. Let us try to understand these things honestly.

As I said in the beginning, it is good to be smart but it is smarter to be good.

Indraneel Banerjee: I heard Dr Cronin's account with great interest, especially the way you enumerated some of the challenges and threats; two interesting ones being, of course which we are all aware of, terrorism and hybrid warfare. You put the challenges saying that these are five of them. Say, in Arabia, if Osama bin Laden put down his five points, he would start with saying that the biggest threat is America or American presence in Arabia. He would say the biggest threat is the presence of Western military in Arabia and in the region. The point I am trying to make is that there are a lot of people with very different perceptions from yours and perhaps just as legitimate. If you are looking for global engagement and trying to work out something globally or trying to understand the dynamics of the world as it is, how do you bridge these different perceptions? Do you say that this is not legitimate or this is legitimate? Or how do you work it out?

Meena Singh Roy: I have a specific question to Dr Patrick or Sir Michael. This is regarding the emerging Sino–Russian strategic cooperation. How do you look at that and the increasing influence of the Shanghai Cooperation Organisation? May I have your views on this?

Patrick Cronin: How you incorporate global perspectives is indeed one of the challenges. Indeed, we are in the business now of creating global networks essentially to respond to globalization. I have not done an adequate job of taking a global perspective but that partly may be because I am reflecting my next position as in about two weeks I go into the US government again. So, I admit that I am already thinking about my next job. I apologize for that. If you had called me a month ago maybe—before I had taken that position—I would have been much more global in my perspective. But more seriously, the issues that you raised, for instance, Osama bin Laden's five points that people should be galvanized by, I think they can be addressed because I think it is part of a false narrative. I think it is legitimate to call parts of that narrative false. While the US power in the Persian Gulf and in the Middle East has sometimes been used for poor purposes and bad purposes, overall I believe the purpose is to try to stabilize a region where there is no stabilizing architecture. There is no overall agreement. We just saw this for the fourth year in a row at Al Manama dialogue for the IISS, a dialogue that was brought about because there was a vacuum of power when it came to how do you bring multilateral actors together and build a regional security architecture. There is no fundamental agreement. We get the Iranians; we get several of the 21 Arab countries together and other outside actors and there is just no agreement whatsoever. So, you are right; there is a multiplicity of perspectives. I appreciate that we should listen to all of them; we should learn from all of them; we should read the original source. But I am giving you my somewhat US-centric viewpoint now at the end of the day.

Briefly on the Sino–Russian cooperation and the Shanghai Co-operation Organization—this is an organization that has four stated objectives. I think to the extent that it lives up to those objectives including those that have been inclusive in dealing with economic issues, in dealing with transnational crime, those are all welcome, welcome by the US, welcome by the West and very good. I think there is some suspicion about it partly because it has not been truly inclusive. That is one of the challenges of a new organization which is to try to do both, not just to network the region, Central Asia, Russia, China, but also to make sure there is enough transparency over the objectives that have been pursued. Some see it rather cynically as something related to energy, for instance. You can look at it through a single prism and be very cynical about something or you can look at something more comprehensively and see it on the round. Perhaps Sir Michael Quinlan has something to add.

Chairman: I think the measure of our complicated subject matter can perhaps be gauged by the number of questions that are asked and I already find four more hands that have been raised.

A.S. Kalkat: First, a small observation for Sir Michael Quinlan. You mentioned a suggestion of making withdrawal from the NPT more difficult or the conditions more difficult. Would it not act as discouragement or a lesser incentive for countries to join the treaty, particularly since your concern about the world seems to be inclined towards the East? As you perhaps will accept, the East is a little charry of treaties. After all, that is the way colonialism came to the East. So, maybe one does not harp too much on treaties. A lot of it works by word because treaties are basically a Western invention. The East was word of mouth in the past. You might like to comment on that.

The other question I have is to Dr Manpreet Sethi. May I please request you to display—you have the computer handy with you—the slide where you talked about proliferation by non-nuclear weapon states. That is where I have a query basically. It is on proliferation by non-nuclear weapon states. There is a mention of Iraq, there, in the second horizontal proliferation; is it an error or is it something that we do not know? You can talk about it later on. It is just a query. Second, if you go back to your other slide, which contains information about accommodation, I suppose this is fine. Would you mind just adding one more reason why India is accommodated? I think it is our democratic values.

Thank you.

I have a question for Dr Bava. Dr Bava, you spoke about waging peace and also using democracy as a political weapon. Do you consider democracy to be a means or an end?

Ummu Salma Bava: General Kalkat, I think you have asked one of the old Platonian questions—process, means and ends. I think democracy is both. It is not just end-based. It is not just where you just ride on a very fine Constitution and say we get there. I think it evolves. As I said, democracy is about conflict resolution within that framework. It churns the ideas about what people want and what they would try to put forward and do it through, what I would say, the ballot and not the bullet. It is about bringing about change in a transformational way in that sense. So, it is a continuous process. It is also a means because it facilitates increasing participation. You would have to look at what is qualitatively there. It is not just about building an institution and, say, once in 5 years or 4 years we go get the little black mark on the finger and then we are done with it.

In terms of moving beyond the electoral to, really, the participatory, you would also have to look at the qualitative aspect of democracy in terms of what is happening if those shifts are taking place.

When I said about waging peace, I took this from what Karen Hughes had said. I think that is interesting to ask. If you are promoting the norm of democracy, then who decides that a state that is not democratic has had adequate time to make that transition? A few select democrats? Would they do that? Is it within the purview of international law because we are talking of governance and as civilized states we think that war is not a policy option? But does war then, in terms of waging peace to create democratic structures, become normatively operative because a few collective democratic states think it is right to do so? I think there is a shift from the notion of the war where we tied it down to the territory, to waging peace, which is tied to the idea. I think that is going to be increasingly the demand. We see a cartelization of ideas. I would like to say, are we going to see willing coalitions or going to stand behind this idea? Because I think it goes back to a point that somebody said from the audience here about—I think Indraneel Banerjee said this—perception determining threat.

My question to you Patrick would be—then how do you build a global perception because geopolitics prevents that kind of bandwagoning to take place? My geography will root me to a particular place and that can never be your geography. That can only be beyond your so-called territorial waters. Would you then come in there to defend it in that sense? Or would others think that this is an invasion of their territorial waters? How do we build a global coalition if our perceptions are going to be determined by our history, by our geography, by our politics and what we endorse as what should be the normative? I think that is going to be a real challenge. We can all say that terrorism is a problem. India has experienced terrorism for very long. We got a new marker with 9/11. Your experience of terrorism is of a different kind. How do we then formulate an overarching, holding definition with which all of us are comfortable, in which all our needs get redressed? I think that is going to be critical. We have endorsed democracy in a very Francis Fukuyama sense. He spoke of the end of history in terms of the world of ideas. I mean it is a danger, if we do not have a mirror to reflect where we are going wrong. Probably that is what the bipolarity in terms of the ideological divide did. It juxtaposed one ideology against the other. You need it. You need it to reinvent yourself. Otherwise, you can be complacent. States can become complacent because you say that this is the epitome of political organization, that democracy is the best, it delivers. Then, we do not look at what happens within if we endorse just electoral institutional democracy.

Manpreet Sethi: Just very briefly, General Kalkat, thanks for your questions. I must say you have been looking at the slides very carefully and noticing.

First, Iraq was not a mistake. Iraq was deliberately put there because that is the point from where we begin to understand horizontal proliferation the way it was understood in the Western construct of the term. Iraq in that sense was the beginning of where they began. It is after that, that the safeguards regime began to be strengthened. It is now beginning to be understood and known that it was a mistake in the sense that the US never had any kind of intelligence on that. Similarly, Iran at some point of time or North Korea might also look like mistakes when they are put on a slide in the future. I mentioned Iraq in that context.

About democratic values, I would say there is no harm if I had put it as one of the last ones. It is a term that has been used more by the US to project the Indo–US Nuclear Cooperation as a thing which is possible with India. One would not discount the fact that the US administrations have been dealing with non-democratic governments with equal enthusiasm. Democratic values in that sense is one of the reasons that they have cited but I do not know if that is the reason that India should be going in for the deal.

B. Raman: We are going to have a panel discussion on terrorism in the afternoon. Without anticipating the discussion, I thought I will make one or two points regarding religion and terrorism because a number of people, and particularly Mr Mardhekar, were very eloquent on that subject. He was asking why is it that when we talk of international terrorism, we focus only on one community? Why always focus only on international Jihadi terrorism? In India, we have terrorism in different places involving members of this community. We had this Al-Umma terrorism in Tamil Nadu. We never tried to create interest in the terrorism, seek the cooperation of other countries in dealing with that because it was purely due to local grievances. We saw it probably as posing a threat to India's national security. Where a terrorism element has global dimensions and poses a threat to international peace and security, the international community as a whole takes interest in that. That is my point number one. But the truth I would like to state is that those people who think they are arguing against creating a division between Muslims or non-Muslims by saying that we should not see it in religious terms, etc., are themselves contributing to creating this division by implying that only the non-Islamic states are concerned about this because only non-Islamic states are affected by terrorism. Islamic states are equally concerned. Saudi Arabia is equally concerned. Egypt is concerned. Algeria is concerned. Croatia is concerned. Jordan is concerned. Malaysia is concerned. Indonesia is concerned. They are much more concerned than any of us. They use much stronger methods to deal with terrorists than the non-Islamic states. For instance, Saudi Arabia has been ruthless in dealing with them. The only thing is that they do not call them Jihadi terrorists as we do. They call them deviants from their religion and they dispose them of ruthlessly. So, we have to constantly keep this in mind that unwittingly, we do not contribute to a polarization between Muslims and non-Muslims by making it appear as if it is only non-Islamic states, non-Muslim states who are talking all the time on this subject or talking all the time on the need for strong action against the terrorists.

Chairman: If there are no further comments or questions from the audience, I would like to request the panellists to individually respond to any of the issues that have been raised during the discussion period. I would also like to mention for the benefit of those members of the audience who were not present last night that one of the issues that we felt is an integral part of any discussion on the strategic shape of the world is with regard to the multilateral trade policy, as we feel you cannot have a really credible or comprehensive discussion on the subject if this issue is ignored. The current state of multilateral trade negotiations on the Doha Development Round has demonstrated how sharply the global community is divided and the scant regard the developed countries have for the developmental needs of the developing world. This subject area was covered last night through a detailed presentation by the Federation of Indian Chambers of Commerce and Industry, highlighting the concerns of India and other developing countries.

I would like to pass the floor on to Patrick.

Patrick Cronin: We start with the comment that really is also may be a defence of Professor Bava on geography in the sense that we can also overstate globalization; we still have to understand, of course, and you did not suggest otherwise Ambassador, the historical context from which we are springing. It is important for the US to understand, when it takes its global view and it conflates everything under global war on terrorism, that the world does not instantly become universally the same just because we say it. It is very important. Having just returned from the Gulf, we were looking for Gulf Cooperation Countries to be more active and positive, constructive participants on the reconstruction of their neighbours in the Gulf, of Iraq. There are many instances where we are looking for significant power simply to do more in the region. That is part of leadership as well.

Other powers around the world, namely Japan, who sometimes have problems acting in their own region, have an easier time acting outside of their region for historical reasons. I would relate this issue to the question on the Taliban and Afghanistan. This is again American penchant not to disaggregate and differentiate on negotiating with the Taliban at least inside Afghanistan. Here, one of the challenges has been the idea that the Taliban has tended to be opportunistic in its gradual recruitment process that has occurred since 2002 when it started to regroup and has now built up a force that is estimated by governments in NATO to be about 20,000 in strength, some private researchers have said it could be 48,000. Whatever the numbers, they are still finite.

It is important to recognize that Taliban opportunistically recruits from various tribes. The challenge for NATO forces, for coalition forces, for Afghan security forces, is not, therefore, to paint an entire tribe from which some people have been recruited as all Taliban. Perhaps, when people refer to negotiating with the Taliban, they are partly referring to the need to differentiate on the ground. Same thing goes with the use of force—that there has to be a recognition that you are trying to protect the population, build economic opportunity and employment, and you are trying to minimize the actual damage. That is, of course, not easily done for armed forces that are sent in the harm's way but it is absolutely paramount when you are fighting a counter-insurgency against the Taliban, of which negotiation is a legitimate part.

Finally, on the question of why the US–India strategic partnership, in particular the nuclear part of that, maybe, Michael Quinlan will have something to say. For me, certainly, Prime Minister Singh's address to the Congress in the US was an eloquent statement about why there is a need and a desire for close partnership and that I think reverberated well in Washington. You have two large economies, vibrant Asian economies that were trying to incorporate into the international system. You have an open democratic system, which appeals to the American values. Yes, you are in the centre of many of the difficult areas of the region. Terrorism and proliferation are obviously high amongst them. The reasons are obvious—that this is a relationship that is long overdue for maturing and strengthening; it will take time; it is a multifaceted relationship. We have much more to learn in the US about India. Hopefully, together the US–India strategic partnership will only strengthen in time and I think the nuclear aspect of it can be a vital part of strengthening the kind of non-proliferation institutions I referred to in my general talk as well as for helping to provide a stable deterrence en route to disarmament.

Michael Quinlan: In response to General Kalkat's question on withdrawal, I agree entirely as a general proposition to have no withdrawal or unduly strong withdrawal provisions in treaties will be a deterrent to signing treaties. But the NPT, I think, is a special case in that it is in fact not a matter of just a treaty but of a powerful global norm. It is not like whether you can withdraw or not, as the US did from the Anti-Ballistic Missile (ABM) bilateral treaty. I maintain, therefore, the view that it is desirable—though we cannot actually as a matter of practicality amend the treaty; that is a can of worms to open—I do believe a clearer set of penalties or disadvantages ought to be patently constructed for the withdrawal.

To Professor Muni, the NPT is indeed discriminatory like the membership of the UN Security Council. It has been that from the start and it held for 40 years. The choice from the start was always that you want this treaty with all its imperfections or none. I must say that that still is the truth. I do not want things to stay that way. I would like to see eventually—if not for my children, at least for my grandchildren—a world without nuclear armoury. But we would not, I think, achieve that or help it forward by jettisoning the current regime.

To Ambassador Ghose on the disarmament question—what should the IISS be doing? I will speak just for myself. It is that the question of eventual complete disarmament has tended to be an empty shouting match between, on the one hand, the pious preachers, whose focus on giving up nuclear weapons is like an international version of die-hard smokers promising on giving up smoking, and the dismissively cynical nihilists. I think we need to try and focus on what would really have to happen, or what conditions would have to be met, how much should be done. The IISS study is not going to try and develop a policy programme or recommendations or even a protection. It is trying to expose and to clarify what would have to be done to achieve a goal, which I think all of us would ultimately endorse.

Since this is my last talk here, can I just pick five other brief points? First, on the UK situation if I could respond a little bit to what Dr Sethi says, what the UK did recently was to take the minimum decision that was necessary to maintain the option of staying in the nuclear weapon business. Had they not taken that decision, they would, in fact, have been saying, 'Our capability will expire by 2020 or thereabouts.' The distinction is no more than maintaining an option for further discussions to be taken, bigger commitments to be made and the field is open.

On the US position, I do not want to speak for the US, indeed even the Bush administration reduced the salience of nuclear weapons. There are no new projects. The RRW project has not been approved by Congress. It is my clear understanding that it is a project that creates no nuclear ability and is designed among other things to make it possible to reduce the stockpile and the reserve held behind it.

Third point is on no-first-use, which is an old chestnut on which I have engaged with some of you before. Currently no-first-use promises or forbids in a condition of stress, as long as you are re-motely controlling nuclear weapons, the idea that your course will be determined by something you have said long ago—this is, I think, quite real. That said, I do think it is the case and perhaps it would be a good idea if both the nuclear weapon powers said that it will be a very bad thing to take recourse to first use, that we think it very unlikely we would ever have to make first use and we will try and conduct both our deployments and our political business in a way which would reduce it to the lowest conceivable level, on the hypothesis that we would ever have to do that. That is, if you like, a policy of no-first-use and that is what in fact the UK has, the US has and I think it would be advantageous if governments said it a bit more clearly and in words, in the way, that I have tried to say it just now.

One last impertinent thought, it is undoubtedly the case that India observes in practice all the norms of the non-proliferation regime. I just wonder whether it might be advantageous politically and helpful for the regime if India formally committed itself as France did for over 20 years when she was not a party to the NPT regime, that though not a party she would in all relevant respects observe the constraints of that regime and perhaps also those of the Missile Technology Control Regime (MTCR) and of the Nuclear Suppliers Group. It would be very good if that encouraged Pakistan to do the same.

Ummu Salma Bava: Ambassador Ghose, I would like to take up the two points that you have raised; one is on coalition building. First, let me flag it by saying that national interest does need to speak. I think that is universally applicable. It is not just for a particular country. Having said that the challenge is that if you are arguing for a normatively built governance structure, we will fall into the trap of the coalition of the winning. There, the challenge would be then in terms of the issue of legitimacy. Will these democratic states by themselves, because they collectively bandwagon, be enough of criteria for legitimizing a particular action? If you look at it in terms of the United Nations (UN) system, I think that will have to be asked. We can jump that entire system and say, 'Well, let us just have ad hoc coalitions which will kind of decide.' I think we will probably get into a Wild West kind of a thing in which a sheriff decides he is today going to take X out or Y out. Who decides then what is going to be endorsed and what is not going to be endorsed? I think that will become very important. I do not think that the jury is back on that on what is to be done. I do not think even for all the criticism of the UN system and of current governments and regimes that we have that we are going to jump that up for simple ad hoc coalitions. I think probably that is where the normative comes in. Yes, if I would speak dual speak that this is its area of interest, its own neighbourhood, it does not want any other, but would want other kinds of normatives. But the difficulty is that I do not think anybody is ready to make that bridge that we simply just sign up to a coalition and can go ahead and do it.

The paradoxes on the other side is that India is in this part of the world, the non-Western democracy and ideally it could not be more closer to the West in terms of the values that it espouses. But that does not automatically lead to a signing up of the Indian flag with the rest. I think the challenge still is that it is not that every idea which is coming out or every cause which is being projected from the global North would be something that India will automatically sign up to. I think that is the challenge. We support the democracy cause. Yes, we do say democracy is an important value; we live it and experience it as we speak. But the process and the ways of building democracy is something India would debate. It is not that India is going to say, 'Well, I am ready to sign on the dotted line.'

On the issue of geography, it is very valid when you say that when we are in a globalized world, what does geography do? I think that is the challenge it poses. I think globalization in one sense creates the opportunity, but it is also leading to this fragmentation in a severe way. I mean it is leading to a major mobility of ideas and capital or people. That is where the geography becomes again political because in the kind of a securitized world we live in now, transfer of people and transfer of ideas are also seen as transfer of threat. We come down to re-protecting the so-called territoriality. There comes geography again. Globalization wants you to escape geography, so to speak. I think your threat once again reinforces that what you are actually trying to protect is the boundary. You return to the kind of modern notion of where the state is. That is the way I would interpret it.

Manpreet Sethi: I will respond very quickly to just two or three points which were raised. First of all, to the very direct question of Ambassador Sahai on Indo–US nuclear cooperation, whether it would make India a global player, my very direct answer is, 'Yes'. One is not discounting the fact that India cannot be a global player without the Indo–US nuclear cooperation but this path does offer you some sort of leverage too; it is like a takeoff point, a flyover over which you could go and reach your destination faster. So, it does act as a force multiplier to that extent, whether it is in terms of economy, energy, or other high end technologies becoming available for India or politically your acceptance as a player in the larger picture of the nuclear thing. On energy, this argument is often made that it is a small thing for energy because all it will lead you to in two or three decades or maybe four decades down the line is only to 11 or 12 per cent of your total energy production. I mean we are discounting the fact that the amount of energy which is being generated in India is really huge. Ten to 11 per cent of that is not a small amount. What are the other options that are available before the country in terms of energy? This is a door that we will be missing to open if we do not do it now. In that sense, I think the Indo–US nuclear cooperation is important. It is one component of a larger strategic relationship but it is an important component.

On the remark made by Professor Muni on whether we should be looking at Chemical Weapons Convention (CWC) and Biological and Toxin Weapons Convention (BTWC) possibly to look at the nuclear-weapons-free world in that whether we can have a nuclear disarmament treaty to that extent, I think at norm-building, CWC and BTWC have done their job. They have set a certain norm into motion. Whether it is agreed to, whether it is accepted, whether it is followed is a different matter, which is why I would like to bring in the point about no-first-use to Sir Michael Quinlan. Well, again no-first-use would set a norm and we could take it upon from there. It is like putting in place a certain parameter, creating a moral repugnance against the use of nuclear weapons. Whether countries actually follow it or not again depends on a host of other factors as we have seen in the case of treaties and export controls.

To the very last point on reduction in numbers, definitely the numbers have reduced in terms of nuclear weapons. But I do not think that is equivalent to the salience of nuclear weapons coming down. The importance of nuclear weapons for deterrence and national security strategies remains just as it was, if not having extended to deterring biological and chemical weapon attacks as well as documents which come out from the US administration to say that regional proliferation will be stopped with nuclear weapons. Those are not positive signs which go out to the international community.

Thank you.

Chairman: It remains for me only to thank the distinguished panellists and the distinguished audience. It has been a very stimulating first session. I am sure the organizers are all going to be quite delighted and hopeful that the other two sessions would be equally stimulating.

Before I invite all of you for lunch, I just thought I would mention that Mr S.K. Singh, former Foreign Secretary of India and presently the Governor of the state of Rajasthan, is trying his best to join us for lunch. In case he is able to do so, he has also agreed that immediately after lunch he would share his thoughts with us on 'The Strategic Shape of the World'. Thereafter, we would move on to the two sessions which are on international terrorism followed by energy security.

Thank you very much. Please join us for lunch.

[Lunch Break]

Address by His Excellency S.K. Singh, Governor of Rajasthan

Speakers
ARIF S. KHAN: Additional Secretary, Public Diplomacy, MEA
S.K. SINGH: Governor of Rajasthan

Arif S. Khan: Excellencies, ladies and gentlemen, it is my great honour and privilege to welcome Governor S.K. Singh. He is a former Foreign Secretary of the Government of India. I am deeply honoured that he has taken the time and the trouble to fly from Jaipur to Delhi to address this conference. I would like to request him to say a few words and give us the value of his experience.

S.K. Singh: Thank you, Arif, for those kind words. When I came here this morning, I knew I would meet a lot of friends from my own MEA and many other friends from the intellectual and academic field. It is a wonderful surprise for me that even among the IISS, there are quite a few of my old friends.

I have been asked to speak to you about 'The Strategic Shape of the World'. I am presently in the state of Rajasthan where most people either live in the past or tend to live in the future. Very few of them are aware of the present. That is something one should say is true of all Governors under the Indian Constitution. They are discouraged from taking a hand in anything concrete. They are encouraged to advise and supervise esoteric activities, nothing concrete, nothing which is of day-to-day relevance. So, I suppose you will find my remarks along these lines, not related to today but to the past, to the future.

The first point I will make is that the changes that took place between 1848 and the death of Queen Victoria in 1901 took a certain time. After Queen Victoria's death, to carry her bier were people who called themselves the 'Eight Emperors': First of them was her own son Edward VII, who called himself the 'Emperor of India'; then the Ottoman Emperor of Turkey; then the Emperor of Ethiopia; the Emperor of Iran; the Habsburg Emperor of Austria; the Emperor of Germany; the Emperor of Czar of all the Russians. Most of these youngsters were her relatives—grand nephews or grandchildren or whatever. As of today, out of those, only one emperor persists, that is the Emperor of Japan. So, that kind of change has taken some time. But changes between the two World Wars, the Treaty of Versailles, in 1945 the founding of the Charter of the UN, took very many fewer years. From 1945 until now, many of the areas covered by the UN Charter—you will excuse my dwelling on that because I have after all spent at least 9 years of my life in various UN centres including Vienna—those articles of the Charter have either been forgotten or are in the process of being changed. We do not see how rapidly the entire Trusteeship Council area has gone away. Nobody thinks about it; nobody thinks about those items on the agenda of Committee 4 of the UN General Assembly with which we used to struggle, and I was there until, pretty nearly, the 1970s.

That is something we must understand. The article on non-interference in internal affairs of sovereign states—whatever their size or shape—has been totally erased. Perhaps the excuse is that human rights considerations have become far more valid. But the human rights considerations being applied even despite the problems created by the Bush Government in smaller countries, smaller powers, are far more; and in larger powers, more powerful powers, far less. Therefore, things have changed. Therefore, international law, the spirit and jurisprudence behind it, is also changing. Having said that, one can say that there have been changes in the understanding of what 'strategic' matters are all about. Strategy used to be geographical; strategy used to be territorial; strategy used to be military oriented, weapons oriented. But the weapons themselves between the end of the World War II or the last year of the World War II and now have changed. Therefore, the change in the strategic understanding of the world today has to be a very different one.

When I was in the UN from 1962 to 1969 in New York, the main considerations before us were food security, the role of the

UN in ensuring that poorer countries, developing countries and the least developed countries got a fair deal. Today, I find from the agenda of the UN and many other bodies that this is an area of diminishing returns. Whatever the pretext, even in terms of use of energy as emissions, etc., it is the rich who want to preserve their bailiwick and blame the developing nations, whatever be the size of the developing nations. That kind of thing as far as developing nations are concerned in Africa is sort of put to basically one large area and that is HIV. Nobody is talking really seriously about erasing their debts or talking in terms of how much exploitation they have already faced in terms of debt repayments. So, I may sound a little old-fashioned in my analysis but these are real problems in Africa. They are real problems in certain countries in the Pacific, certain countries in the Caribbean, smaller economies. But these are the people who are close to us, India, and I expect also to China.

Therefore, whatever productivity China has developed—and we congratulate them for that—and whatever productivity we have developed, whatever our requirements of energy are today, how we tackle our own respective problems of infrastructure building, all this is tied up with the rest of the world. Similarly, I would say that all of you know perfectly well that some kind of terrorist activity must have been there in ancient times also. Actually, all kinds of stories that one hears in the Greek history or Roman history or Indian history or Chinese history in terms of court intrigues, etc., were of a terrorist nature. As far as India is concerned, our Independence came along with our partition. Partition was a period of bloodshed. This period of bloodshed was followed by a feeling that one major neighbour in the North-East could keep us on the defence and there were—I have lived and worked in the North-East recently and therefore, that is quite higher up in my consciousness—people whom a big neighbour could buy with money, with influence and with a few weapons. That happened.

Similarly, our western neighbour did put us under considerable pressure in Kashmir. That also resulted in a state-aided terrorist activity for a very long period. So, to us, terrorism is not new. Even as an independent nation, India has faced terrorism all these 60 years of Independence. But whenever we raised the point of terrorism by whatever name, whatever nomenclature, we were told to mind our own business and to improve our communal relations, our administration of justice, to have a more efficient police, etc. But since

9/11, the whole world's attitude towards terrorism has changed. I may have somewhat different impressions of terrorism for terrorism as an abstract entity is a global enemy. Terrorism is fairly abstract. Terrorists are concrete and terrorists can be handled in a concrete way. Anyway, all that is there in the Convention, which India had drafted and which was hanging on the walls of the UN for several years before anyone got prepared even to read the conditionalities there. So, I would not go into that.

The last point I would make is that for many of my ilk in India, one of the most important things in which China and India can and perhaps should be cooperating and collaborating is the Indian Ocean. From the point of view of energy delivery to us and energy delivery, bulk of it, to them, it should be important for us to develop our patrolling of the Indian Ocean, our keeping the sanctity of the Indian Ocean, keeping it an area of peace one way or the other. But it is not happening so far because, perhaps, the past is impinging on the present and therefore, the future and the 'border' kind of activities or rather the 'river mouth' kind of activities are continuing. But I think that if they think in terms of better management of their economy and we think the same way ourselves, we should be cooperating and collaborating in some years' time.

I believe you have already discussed the non-proliferation issue quite threadbare. So, I would not repeat anything that might have been said. These were my basic thoughts on what conditions India's attitudes towards strategy today and tomorrow.

Thank you.

Arif S. Khan: Thank you, Governor, for your most informative address. Since you arrived at lunch, I noticed that you seem to know almost all the delegates who have come from the IISS. Nevertheless, I would like to mention that we have with us Dr Patrick Cronin, Sir Michael Quinlan, Sir Hillary Synnott, Nigel Inkster, Jack Gill and, of course, Rahul Roy-Chaudhury, who has been the backbone from the IISS side in organizing this meeting and the earlier one between our two countries. I would like to thank all of them. I hope that the deliberations, which will be published, would take us forward in many ways in our dialogue.

Thank you.

Session II

International Terrorism

Speakers
CHAIRMAN: Amit Dasgupta, Joint Secretary, Public Diplomacy MEA
B. RAMAN: Indian Lead Speaker
NIGEL INKSTER: IISS Lead Speaker
DR NAVNITA CHADHA BEHERA: Indian Discussant
SIR HILLARY SYNNOTT: IISS Discussant
NAGMA M. MALLIK: Director, MEA
SAVITRI KUNADI: Former Ambassador
ISHRAT AZIZ: Former Ambassador
UMMU SALMA BAVA: Professor of European Studies, Jawaharlal Nehru University, New Delhi
NIGAM PRAKASH: Former Ambassador
DR PRADEEP SHARMA: Research Officer, Thai Embassy, New Delhi

Chairman: This morning, we had a general session, which then moved on to the micro topic of nuclear proliferation. The first session post lunch is on an issue of global concern: international terrorism. As the Honourable Governor said just a few minutes ago, India has been the victim of terrorism for several years, indeed, since our Independence 60 years ago. It is an irony of fate that it required a singular event, 9/11, to start focusing attention on the fact that terrorism is multi-locational, it knows no borders and that anyone can be hit at any time.

We have very distinguished panellists with us today. The rules of the game remain the same. There are two lead speakers and two discussants. Each lead speaker speaks for around 15 minutes and discussant for 10 minutes. Then, we will open the floor for discussions or questions. We have about an hour and a half to go through this particular subject.

The lead speaker who would start the deliberations would be someone all of us in India know, Mr B. Raman, who retired as Additional Secretary a few years ago. Mr Raman, the floor is yours.

B. Raman: I feel greatly honoured to be in front of all of you and to be a part of this seminar and exchange some ideas. We do not want...my assessment of the state of international terrorism today. ...wrap it within the 15 minutes allotted to me but with much of...I will be happy to deal with them. One is to mix the old and the new terrorism: the old targeted killing of select individuals or groups, and the new is mass casualty, mass destruction. It is so different to impact on their actions...of international Jihadi terrorism are not contemporary. They rise from their perception of Islamic history. They create a Caliphate and restore what they describe as historic Muslim lands not with modern Muslims back to the Ulema.

...such as unemployment, poverty, social injustice, etc., do not apply. Bin Laden does not even talk of such issues. He talks only of historic issues such as wrongs committed to Muslims over the course of history, focus on Muslims as a community, Islam as a religion and not as Muslims as individuals. For him, the importance of individual Muslims is as suicide bombers and nothing else.

Coming to the defining characteristics—it is not a classical pattern. Many operatives come from affluent or well-to-do families. Many operatives are educated and well-employed types. There is no distinction between combatants and non-combatants or between men, women and children. Jihadi terrorists have killed more innocent civilians in the world than all other terrorist groups put together, mainly using explosives. They are prepared to use any means of mass casualties, even weapons of mass destruction. Tactical objectives are mainly acts of reprisal for Palestine, Iraq, Afghanistan and other perceived wrongs done to Muslims. Strategic objectives are—massive economic disruption, target transport, tourism, maritime trade, oil and gas production and transportation, and critical infrastructure. Previously, they wanted to target only oil, gas transportation, but not production facilities located in Muslim lands. Latest instructions are to target production facilities too, even if they are located in Muslim lands.

Continuing with the defining characteristics, it has acquired a self-sustaining momentum. Elimination of iconic leaders has had no impact on the momentum. It is rank and file terrorism. Individual elements are capable of autonomous operations even in the absence of an iconic leader. From the pre-9/11 hierarchical set-up, Al Qaeda has evolved into a set-up of concentric circles with each circle representing a local initiative with the operational command and control at the centre available for guidance when required, but not otherwise intervening.

Defining characteristics again: importance of martyrdom operations, that is, suicide terrorism; importance of two-pronged operation—martyrdom and *intifada* tactics for those who are disinclined to commit martyrdom; affiliation to organizations is not obligatory. A Muslim is free to wage Jihad either as a member of an organization or as an individual—the Jundullah, or soldiers of Allah, phenomenon—individual Muslims not belonging to any organization seeking martyrdom. The Jundullah phenomenon is on the increase in Pakistan. There were four suicide terrorist attacks in Pakistan in 2002, two in 2003, five in 2004, two in 2005, six in 2006 and there is a manifold increase to 41 till 28 October 2007. Total deaths in Pakistan—796. Most of the cases of 2006 and 2007 remain undetected. Suspected reason—individuals with no previous history and with no previous organizational affiliation involved.

Sunni bomb versus Shia bomb. Support for Sunni bomb. Support for Pakistan acquiring a nuclear weapon. Silence Iran's right to acquire an atom bomb. Loud condemnation of US intervention in Afghanistan and Iraq. Muted reactions to reported US plans to intervene in Iran.

Bin Laden's so-called International Islamic Front is in fact an international Wahabi front against Christians, Jewish people, Hindus and Shias and their Muslim state supporters. Muslims killing Muslims, we see it in Iraq, we see it in Pakistan, we see it in Afghanistan, we have been seeing it in India.

What has been the ground situation today? Tentative signs of improvement in Iraq—US military deaths dropped by 63 per cent from 101 in June 2007 to 37 in November 2007. The US military deaths in Baghdad also decreased by 78 per cent from 40 in June 2007 to 9 in November 2007. Deaths from Improvised Explosive Devices continue to be the highest incident type but dropped by 72 per cent since this summer. Iraqi civilian deaths from war-related violence dropped by 56 per cent over the past 6 months from at least 1,640 in June 2007 to 718 in November 2007.

There is no significant sign of improvement in Afghanistan. The neo-Taliban and its associates caused 346 civilian deaths, while NATO and Afghan forces caused 337 civilian deaths till the end of October. Military fatalities since 7 October 2001, when Operation Enduring Freedom started: US 469, Britain 85, Canada 73, Spain 23, Germany 26, other nations 66, total 742. According to a UN study, the number of suicide bombings in Afghanistan increased from 17 in 2005 to 123 in 2006, and touched 103 till 31 August 2007. It is likely to exceed last year's figures by the end of this year (2007). A unique situation—a mix of insurgency and terrorism, of conventional and non-conventional modus operandi, neo-Taliban in the driving seat and not Al Qaeda.

Worrisome deterioration in Pakistan, South and North Waziristan, in Federally Administered Tribal Areas (FATA) and a de facto control of a hotchpotch of terrorists—Al Qaeda, local tribal groups close to neo-Taliban, Uzbek and Pakistani Jihadi groups, individual Jundullahs spreading to areas outside FATA. Four attacks in Rawalpindi—two against Army and two against the Inter Services Intelligence (ISI), one against the Air Force in Sargodha, and one against the US-trained Special Services Group in Tarbela, many attacks on the police in the North-West Frontier Province.

In India, there is improvement in Jammu and Kashmir. Civilian fatalities in Jammu and Kashmir declined from 521 in 2005 to 349 in 2006 and 162 till 5 December 2007. Security forces fatalities in Jammu and Kashmir declined from 208 in 2005 to 168 in 2006 and 120 till 5 December 2007. Worrisome situation prevails in the rest of India outside Jammu and Kashmir. Three terrorist strikes in 2005, three in 2006 and eight so far this year. Of the attacks since 2005, three were on Muslim places of worship, three on legal community, two on transport, two on Hindu holy places, three in public places, and one on a meeting of scientists. One attack was with a gun and the remaining 13 were with Improvised Explosive Devices. Estimated fatalities—300. More use of handheld weapons in Jammu and Kashmir. More use of explosives in the rest of India.

Operational control is still done by Pakistan's ISI but the command and control increasingly operates from Bangladesh after Pervez Musharraf assured India in January 2004 that he will not allow any territory under Pakistani control to be used by anti-India terrorists. ISI continues to use pro-Al Qaeda terrorists against India

and the neo-Taliban against Afghanistan. For the rest of the world—
sporadic as ever but lethal as ever. New likely danger spots: Germany,
Denmark, France and Canada. Larger reservoir of operatives than
before 9/11, Arabs, Pakistanis in Pakistan and diaspora, Uzbeks and
Indian Muslims.

Pakistan: how do you look to the future?

Senior army leadership sees no alternative to Musharraf in the driving
seat of national security, his operational manoeuvrability diluted
by his making General Ashfaq Pervez Kiyani as the Chief of the
Army Staff. When Musharraf chaired the Corps Commanders'
meetings as the Chief of the Army Staff, his officers avoided openly
voicing their reservations over his policies. They will be less inhibited
in voicing their reservations in meetings chaired by Kiyani. He has
to take them into consideration. Upsurge and tribal anger after com-
mando action in Islamabad's Lal Masjid continues. Instability in the
Pashtun belt will continue in the short and medium terms. Upsurge
and terrorism will continue; tribal areas will remain ungovernable,
non-tribal areas will keep bleeding intermittently. The state of affairs
will continue till the Jihadi sanctuaries and training infrastructure
in Pakistani territory are effectively eliminated, whether they are
directed against India, Afghanistan or the US.

There is no end to Jihadi terrorism originating from Pakistani
territory so long as the US keeps giving Musharraf cause for belief that
it will close its eyes to his Jihadi terrorism against India provided he
helps it in preventing another 9/11 terrorist strike in the US territory.
The nightmare over Pakistan's nuclear weapons and rogue nuclear
scientists will continue haunting the US so long as Jihad continues
to flourish in Pakistani territory. If there is another 9/11 in the US
and if it involves the use of weapons of mass destruction, it would
originate from Pakistani territory.

Counter-terrorism: what are the inadequacies?

Action against funding is largely focused on formal channels such
as banking. Informal channels such as *havala*, use of business companies and stock markets, etc., are not effectively addressed by the

international community. Ineffective narcotics control; ineffective action against sanctuaries and training infrastructure in Pakistan and Bangladesh; ineffective psychological warfare; terrorists make better use of soft power against state actors and vice versa. International cooperation is confined to intelligence sharing; there are very little joint operations; reluctance to share appropriate technologies.

Before April 2006, the global Jihad was projected as directed against the Christians and the Jewish people only. Since 2002, there has been a flow of Indian Muslim volunteers from India and the Gulf to the four Pakistani organizations mentioned earlier. Al Qaeda had definitely used one Muslim of Indian origin in the UK for intelligence collection and target selection in the US. One other Indian Muslim in the UK was suspected of association with the London bombers of July 2005. The third Indian Muslim who was involved in the Glasgow incident of June 2007 was acting individually. There was no evident organizational affiliation. Khalid Sheikh Mohammad, who is in US custody, reportedly told the US interrogators that Al Qaeda wanted to strike the Israeli Embassy in New Delhi but could not. It is assessed that Al Qaeda will continue to leave the operations against Indian targets to its Pakistani associates but will keep looking for an opportunity to strike at US and Israeli targets in India.

India: difficulties faced

Wide geographic spread, involves coordination among many states; it is gypsy-like terrorism; they [terrorists] keep shifting their areas of operation from state to state. Unestimated but large population of illegal migrants from Bangladesh all over the country—they provide the logistics support. Reluctance of political leadership to give additional powers to the police on par with those in the West. Reluctance to act against illegal Bangladeshi migrants.

General observations

Conventional root causes and hearts and minds approaches do not apply to Jihadi terrorism since its so-called grievances are not contemporary. It has to be effectively neutralized through security measures—defensive and offensive. Over-militarized US counter-terrorism methods with disproportionate use of force through air and artillery strikes is adding to Muslim anger and driving more Muslims into the waiting arms of Al Qaeda. There is a need for mid-course corrections in the US' counter-terrorism policy. India's non-military approach is worthy of emulation. The Indian model includes police, the weapon of first resort in counter-terrorism, and army, the weapon of first resort in countering infiltration in border areas.

Pakistan has been the most important breeding ground of inter-national Jihadi terrorism. Bangladesh is the next. The armies in both countries have not sincerely addressed this menace. There is a need for new leadership in both countries, which will tackle the menace sincerely and effectively. Intelligence has improved but is still facing difficulties in penetrating non-state actors. Physical security has improved but it is still facing difficulties against suicide terrorism and attacks on soft targets. There has been no breakthrough in developing technologies to detect suicide bombers from a distance. There has been a dramatic deterioration in the morale of the police in Pakistan under Musharraf.

The last is a thought for the road as you leave this place. The US is caught in a vicious circle in Pakistan. The more it supports Musharraf, the more would be the anti-Musharraf, anti-US anger. The more the anti-US, anti-Musharraf anger, the more the Jihadi terrorism. The more the Jihadi terrorism, the more the dangers of Pakistani nuclear arsenal falling into the hands of Jihadi terrorists. The more the dangers of the Pakistani nuclear arsenal falling into the hands of Jihadi terrorists, the more the US will support Musharraf. So long as the US does not get out of this vicious circle, its troops will con-tinue bleeding in this region. There is no returning home before Christmas for the US troops in this region, not for this Christmas, not for many Christmases for years to come until their policy towards Pakistan changes.

Thank you very much and wish you all a very happy Christmas.

Chairman: The next speaker is Mr Nigel Inkster, who has served several years with the British Intelligence. He is currently the Director of Transnational Threats and Political Risk at the IISS.

Nigel Inkster: Thank you speakers from the IISS. I should make it clear that I am no longer with the British Government. I left Her Majesty's employ at the end of last year and as far as I am aware, she is not reportedly suffering from withdrawal symptoms. So, my views are my own.

The previous speaker, Mr Raman, has already defined the debate in terms of international Jihadist terrorism or Islamist terrorism, call it what you will. Of course, that is not the only problem of transnational terrorism that we might face. I think that looking to the future, an area that we will need to keep a close eye on, is the potential evolution of some extreme environmentalist groups into this area. We have already seen one example of such people, so to speak, crossing the species barrier in New Zealand and engaging in what seemed to be some kind of preliminary terrorist training. But that is something for the future, it has not really happened yet. I, therefore, think it is sensible and indeed incumbent upon us to concentrate on the key area. which is the Islamist, Jihadist terrorism.

Mr Raman described this now as consisting of concentric circles. I would not dissent that analysis; I would characterize it slightly differently, dividing it into three areas. First, what we might call old or core Al Qaeda—the people around Osama bin Laden and Ayman al-Zawahiri in the tribal areas of Pakistan. Second, what I would call the affiliates—organizations like the people who conducted the bombing in Algiers the day before yesterday, Al Qaeda in the Maghreb, an organization that has sworn formal allegiance to Osama bin Laden. The third, what I would call the independents, which is a particular problem we are encountering in Western Europe. These groups of young men and women, let us not forget the women, can be described as self-radicalizing largely through the medium of the Internet, self-forming as groups—there are no external influences that bring them together—and operating almost exclusively outside any formal Islamic structure. These groups do not form themselves in the margins of the mosques. They form themselves in schools, in gymnasiums and in other areas of informal activity.

When one is looking at terrorism, it only makes sense if one can very quickly move from the conceptual to provide a degree of detail and that is what I would like to do—try and put some flesh on the bones of this. What I hope to do in the course of the next few minutes, it is a big task I realize, is look at what Al Qaeda has become, question how effective the universalist Al Qaeda message is, and speculate about how this phenomenon might develop and change over the coming years. Let me look at what is happening with Al Qaeda Central as described by Mr Raman. They are there largely in the federally administered tribal areas of Pakistan. We do not know exactly where Osama bin Laden is. If I were a betting man I would guess that Aiman al-Zawahiri is probably in Bajaur, the northern most tribal agency, for at least 60 per cent of the time.

You have, as was described earlier, a rich brew of Al Qaeda consisting predominantly of Arabs of Egyptian, Libyan and Maghrebi origin. You have other foreign fighters like the Uzbeks and the Chechens who are aligned with Al Qaeda but are formally separate. You have the Taliban, which covers a multitude of sins from people like supporters of Mullah Omar to young Pashtuns who are simply working for whoever will pay them most. You have what I would charitably describe as the remnants of Kashmiri groups and then you have the tribes in the federally administered tribal areas, many of whom have significant military capabilities, armies of 15,000–20,000 people with quite significant levels of armament. So, that is a rich brew.

How does Al Qaeda operate? There is a Shariah Council. I doubt whether it formally meets as a Shariah Council, as a Board of Directors. But it does seem to be able to operate rather like a Shariah Council. There is communication and there are designated responsibilities. We can say with confidence, for example, that someone like Abu Obeida al-Masri is the head of Al Qaeda's external operations. We know who is responsible for communications. They appear to be able to communicate and to some degree coordinate between themselves, though I suspect that that coordination is pretty loose.

Higher levels of security

Messages to and from Al Qaeda Central typically now go through the couriers who have memorized texts and pass through several cut-outs before they reach their intended destination. Al Qaeda has largely given up on the use of electronic communications having realized how vulnerable to interception they are.

Training

Jihadist training is one of the phenomenon that particularly concerns people in the West and the extent to which the self-forming groups I described earlier have managed to get to the remote areas [of Pakistan]—where the Pakistani Government's writ, let us not forget, has never really run—and undertake Jihadist training. We have seen this happening in the UK, we have seen it happening in Germany and it is starting to happen, I think, more widely. In most cases, it is the radicalized groups who take the initiative rather than Al Qaeda reaching out but it is to some degree a two-way process.

The training is actually not carried out by the Al Qaeda. It is subcontracted to Kashmiri groups, to organizations like the Islamic Jihadi Union that trained the people who undertook the Frankfurt and Ramstein attempts earlier this summer. Al Qaeda is able to sit back and cherry-pick those operatives who look most promising for their purposes. We should, however, note that the quality and level of training on offer in these camps is not that great. It is basic, as we might say in England in the 21st century, bog standard Jihadi training—how to use Kalashnikov, how to make bombs using pretty basic household ingredients. There is nothing very sophisticated about it. What we have seen certainly in Western Europe is what I would call a Jihadi cycle, people going to these camps in the spring, getting trained, coming back to the West during the summer and then trying to put together an operation.

A degree of direction from the centre and big ambitions—if we look at the last two major plots that were disrupted in Western Europe, the so-called airlines plot which was intended to involve up to 10 airliners exploding over mid Atlantic but was also to have been followed by a second wave of attacks against critical national infrastructure, and the bombings that were disrupted this year in Germany, Ramstein airbase and Frankfurt airport. Big ambition, limited capability to implement this ambition, and the quality of people involved not that high, clearly operating under enormous pressure from Al Qaeda to get a result quickly, to get some points up on the board—that was very much the impression that one got last year with the airliner plot—leading people to move too quickly, be careless and get caught. We have seen also that there is no inclination or no ability on the part of these people to move up the value chain. Much talk of the use of Condition Based Risk Management (CBRM)—little in the way of convincing activity. Even Dhiren Barot to whom Mr Raman referred obliquely, who was a very competent operative—I would certainly have given him an A minus for some of those reconnaissances he conducted in New York and London—had some, frankly, mad ideas about how you might construct a radiological dispersal device using americium from smoke alarms. Well, you would need to fill this room with smoke alarms before you came close to achieving that kind of outcome. So, I think there is still a significant gap between ambition and capability here. I want to come back to that.

Moving very quickly on to what I call the subsidiaries, including Iraq which was also mentioned. We have seen an organization evolve from so-called Al Qaeda but formally now known as the Islamic State of Iraq, not to be confused with that fine body of men just across the border in Pakistan. The focus of that group has basically been on doing two things—getting control over the Sunnis and attacking the Shia. There have been considerable tensions in that agenda with Al Qaeda as we saw from intercepted correspondence. We have seen the population turn against the Islamic State of Iraq in some areas, notably Anbar, an exercise, which I doubt can be replicated elsewhere because the tribal structure is intact in Anbar in ways which it is not in the rest of Iraq. Al Qaeda, let us call it that, has not gone away—it simply moved North to Mosul where the action is about to be. We have not seen blow back from Iraq, which was something people were very worried about. We have not seen people go, become hardened Jihadis and then come back to other places and carry on their trade, mainly because the only foreign fighters who are welcome in Iraq are suicide bombers and tend to travel on a one-way ticket. I think blowback is something for the Middle-Eastern States to worry about in the future, no question about it. We certainly cannot dismiss Iraq as a mobilizing factor for the radicalization process. It is clearly important.

Ultimately, I would argue that the relations between the Al Qaeda in Iraq and—also, I think this is true of organizations in Algeria, in Saudi Arabia, Al Qaeda in the Saudi peninsula, which has shown itself active and persistent—organizations like Libyan Islamic Fighting Group are marriages of convenience. I think that there are at work powerful centripetal forces working against Osama bin Laden, as a, what one might term, universalist agenda. Patrick, at the beginning of the proceedings, referred to the need to deconstruct this phenomenon. I think this is an important part of the general approach in pushing back against what we call a single narrative—the idea that everything that happens is boiled down to one simple explanation of the West persecuting Muslims—and to actually address the real issues at a regional and sub-regional level.

I would like to touch very briefly on communication and then move quickly to one or two conclusions if I can. Obviously, one characteristic of Al Qaeda is the way in which they have been able to disseminate their ideology very effectively. Al Sahab, the Al Qaeda media organization, appears to be able to function very efficiently, and turn around stories very quickly. They do not need much in the way of technology to do that. They need an electricity supply, a laptop, a video camera and a distribution network. That is really all they need. They do have some quite sophisticated and slick media spokesmen including Mr Adam Gadahn, a former California surfer dude, who is now the main Al Qaeda spokesman for the West. Exploitation of the Internet is an important part of the process—using sites which were often password protected and chat rooms which appear to be autonomous spaces where you are radicalizing young men and women, [where they] can ostensibly exchange ideas freely; but often it turns out that they are being clandestinely guided in a particular direction. It may be that certainly in the case of Western Europe, there are as few as half a dozen key chat rooms involved in this process. One of the biggest difficulties again that has been referred to is the collective problem of rebutting the single narrative, creating an effective media operation that pushes back against this message. That I think is one of the biggest challenges that we face.

Where is all this going? What conclusions can we draw? After 6 years of the global war on terror, a concept we in the Institute do not particularly subscribe to, Al Qaeda has not been destroyed. In fact, they have shown themselves to be remarkably resilient. They have been able to conceive and plan operations with significant reach and ambition even if their success has not been that great. Like all guerrilla groups, they win by not losing and by staying alive in the hope of better times to come. As I said before, the quality of the operations that they currently undertake obviously is deadly but they have not moved up the value chain. What would it take for them to do that? I think the short answer is a step change in the quality of the leadership of the groups that are actually implementing these plans and that has not yet happened.

What are the possibilities? One possibility that I do not think we can altogether discount is that 2 or 3 years down the track we will look back and conclude that 9/11 was the high watermark for all of this, and that Al Qaeda have not been able to move ahead and make a greater impact. Another alternative is that if politics in the Middle-East becomes defined by the Sunni–Shia split, Al Qaeda transmogrifies into a kind of shock force for the Sunni rather as Osama bin Laden had in mind at the time of the first Gulf War—you may/would remember that he wanted to sort out Saddam Hussein, so the Americans would not have to be invited in to do it—that is something we should not overlook. Neither of these would be particularly comforting options. Even if 9/11 is the high watermark for Al Qaeda, that does not mean that we cannot expect a great deal of turbulence and trouble in the future. This sort of incidents that have been described can carry on at the current sort of level for a long time.

We need to remember one thing that terrorists do not need to be that good; they need to be lucky. If one of the operations that has been frustrated over the last 2 or 3 years had come to fruition, the terms of this debate would look very different from the way that they look now, particularly if something had been done which caused serious economic destruction of the 9/11 variety. Common prudence dictates that we have to proceed on the basis that risk exists and do what we can to avert it.

One final point is international cooperation. I certainly dissent from the point that joint operations have not taken off. That does not accord with my personal experience, having been directly re-sponsible for working on this for the last 2 or 3 years. I think it is happening more and more. Intelligence services, police forces, and security services are increasingly speaking the same language, exchanging information better. We do have some rules of the road. There are some common definitions. The EU has a common def-inition of terrorism. The UN has done a lot of work to pull together thinking and agreement on terrorism and appropriate political and police responses to terrorism.

We are making some headway. Clearly, there is always going to be more that we need to do. But I do not think that we can afford to allow ourselves to be deterred by the fact that more has not been done. At the end of the day the international community in my view can only ever really create an enabling environment. It has to be the nation states, which are the units we have to work with, which actually do the delivery.

Thank you.

Chairman: Could I now invite Dr Navnita Chadha Behera, Professor of the Nelson Mandela Centre for Peace and Conflict Resolution, Jamia Millia Islamia, to make her presentation.

Navnita Chadha Behera: Thank you Mr Chairman for giving me this opportunity to be here. I think my task has been made much easier by the preceding two very comprehensive presentations. I will try my best to be brief and to perhaps highlight some of the points that have not yet come into the debate. There may be a few points of overlap. I will try to avoid that.

Taking off from what Mr Raman had said in terms of the defining characteristics of the mix of new terrorism and the old terrorism, what struck me was also a factor that in the olden days too, when terrorism was used as a technique, it was always used in a combination of a revolutionary warfare. The targets were always very political, very military. I think what distinguishes the new terrorism is the indiscriminate targeting in terms of civilians. There is no longer a premium on choosing a high value economic and political target. By sheer mass killing of innocent civilians, they are able to get the attention that they want. That is something we are witnessing more and more. But 9/11, I agree with the previous speakers, has been a bit of a watershed point in reinforcing the point that terrorism knows no borders. The earlier phenomenon that it was a problem confined to some states was really put paid to with 9/11 that no state is immune to the threat of terrorism. But the trouble I find is that after fighting the so-called global war on terror for 6 years, there still seems to be a little lack of clarity in terms of its objectives. It is partly becoming an open-ended game perhaps because there is an ever-growing list of who your enemy is. There is no fixed list. There is no fixed list even of the terror groups. New ones keep getting added on a daily basis. Is it still primarily against Al Qaeda or is it Al Qaeda and its various mutants and variants?

It is partly, to my mind, because US's original approach after 9/11 of [addressing] Al Qaeda first and other groups later—as in HuJI or Lashkar-e-Tayyaba or Jaish-e-Mohammed—giving them secondary priority; that is one policy that one must acknowledge has not delivered. Al Qaeda has not really been decimated. If you look at the incidents of international terrorism—be it in Iraq, in Afghanistan, in Pakistan, in India, elsewhere—there is growing albeit grudging recognition that Al Qaeda has not only just mutated into several smaller number of groups but they are all swimming in the same tank. So, you cannot really distinguish Al Qaeda from others. You cannot target only Al Qaeda and say that others will be addressed later. You have to pretty much widen your horizons because that is the world view that they are working in. The terrorist groups are working in a coalition of like-mindedness where they share their resources, they share their targets, they share their weapons, funding, everything. One needs to acknowledge that. The net effect is that you are not targeting only Al Qaeda but also focusing your attention on others.

The second point is that the US policy on fighting terror is still a matter at a fairly macro political level debate that whether fighting terror is a goal in itself or is it still in some policy-making circles considered as an instrument of policy for reordering the world. The reason I am mentioning this point is because sometimes these two objectives are at cross purposes. If we are only fighting the war against terror, then Iraq and Iran do not really fall in place. They were really not on high priority in terms of being bases of Al Qaeda or WMDs. Choice of some of these targets leaves room open for debate as to where we are headed.

The third point is, taking up from what Mr Raman mentioned, that the old theory of poorly, badly governing society becoming the breeding grounds of terror is not really working very effectively any more. The recent attacks including the 7/7 attack and the Glasgow attack show that we are witnessing new recruits to the cause of terrorism, which is very clearly from an educated intelligentsia with no previous records of having been involved in any kind of criminal activities whatsoever. So, the motivations here are entirely different. This calls into question also another old familiar theory of addressing the root causes of terrorism and eliminating it. What I am trying to say is that it should not be misconstrued as not wanting or not needing to understand the political dynamics of what is causing terrorism, but we must understand and recognize that terrorism has acquired an independent logic and a momentum of its own. In many circles, it is becoming an end in itself. For example, in Kashmir, where we have done a lot of work, if the original objective of some of these groups like Lashkar-e-Tayyaba, JeM and HuJI was to gain Kashmir for Pakistan, over a period of years this policy of bleeding India through thousands of cuts has become an end in itself.

I came across this very interesting concept in the book *The Quranic Concept of War* by General S.K. Malik, who holds a very high position in the Pakistan army. He says so much in his book that the objective of fighting Jihad is just to cause terror in the hearts of the enemy. That itself is a goal. It is not an instrument of policy any more to achieve any identifiable, concrete, political end. It is becoming an end in itself.

The third general point that I wanted to make was in terms of the counter-terror measures that we are taking. Mr Raman has given us a very detailed exposition of what is the range of physical security measures that are required and necessary as a safeguard device. But in adopting them I think the Governments also need to make sure that they do not turn unethical. That is because in the long term, if the states were to use their legal framework for unethical ends, it will boomerang on us because you lose your high moral ground there. It is happening in several ways. For example, in the American case, it is not providing judicial access. We are in the legal framework hearing new categories of illegal combatants and the Geneva Convention not applying to them. Also in another sector, somewhat unrelated, they should not be used somewhat as a backdoor entry for reshaping the immigration policies—which is happening—which are very highly restrictive. But what is often happening also is that a lot of genuine travellers are also suspected of terrorism. This is increasingly happening, of course, for oriental countries like Asia and the Arab world. It can be counter-productive because what it is doing is also alienating several elements of the civil society intelligentsia against some of these policies because consciously or unconsciously, they are creating a divide between the communities as to who is more likely to be suspected and who is above the cause of suspicion.

Another important factor to my mind is to keep a balance between the political and security measures in fighting terror. That, to my mind, is the real key. It is an appropriate mix of strategies which is needed. If you were to look at some of the individual country's measures, we often find it lacking. For example, in the US, there is a really very heavy focus on homeland security measures but these are what I would call in the range of physical security measures, protecting the country in terms of tightening the border controls, tightening the immigration policy, identifying who is entering the country and who is leaving the country. But it still has not come to terms with the reality that some of its policy actions, for example, in Iraq and otherwise, in Afghanistan, are also providing ground for further recruits to the cause of terror.

In Britain and France and other parts of Europe, I find the challenge is somewhat different. It is in terms more of integrating their migrant communities into their mainstream because these are really mixed with their own populations. Some of these elements, as the Glasgow attack and the 7/7 attack show, are getting radicalized. Mr Nigel pointed out very effectively that not all of them are necessarily linked to the Islamic cause but they are getting radicalized, getting wedded to the cause of using terror as a technique to drive home their point.

In the region, I find the dynamics slightly different because we have very clearly earmarked communities. Whether it is Kashmiris in India, Tamils in Sri Lanka and Baluch in Pakistan, the relative dividing lines are much clearer. The political class in these countries needs to adopt a different approach in terms of reaching a political understanding or a political reconciliation with the alienated segments of its own society. What I find interesting is that a country like India has done very effectively, for example, in the North-East and Punjab. The resilience I find of Indian democracy system is of being able to share power with the sections that are unhappy. That removes the political raison d'être for communities to resort to means of violence. But it does fall woefully short in terms of the physical security measures. You know our borders are still very lax. The issue of differentiating between citizens and aliens does get politicized. Mr Raman pointed out the case of Bangladeshi immigrants in the North-East region. It is a real problem, which does need to be addressed.

One last point about the region is that I think there is a lull in violence in Kashmir. But I think it should not be mistaken as a sign of victory or having defeated terrorism. Terrorism has not really gone anywhere. It is just keeping a low profile, at least partially at the moment because Pakistani armed forces are fully engaged on its eastern borders in North and South Waziristan. To assume that they will never turn back on to the western side would be a mistake. If we have to tackle the problem of terrorism in the region at least, we have to go up to its infrastructure and capability. So long as the infrastructure remains, the tap can always be turned on. That is something I think is a real problem that we have.

The targets of terrorist attacks in India are no longer really confined to Kashmir. If you look at the trend in the past couple of years—whether it is Mumbai, Lucknow or Bangalore—it is spreading right throughout India. Also in terms of their entry points, routes, Nepal and Bangladesh are being used more. Their choice of targets, their personnel who are able to get away after conducting terrorist attacks back into other countries, are getting much more diversified. To that extent, I think the battle is becoming a lot more complex and a lot more difficult to fight.

Thank you.

Chairman: Thank you very much Navnita. I would now like to invite Sir Hillary Synnott to make his presentation as the discussant. Sir Hillary, prior to joining the British Diplomatic Service, was for 11 years with the Royal Navy where he was a submariner. During his several assignments in the British Foreign Office, from 1993 to 1996, he was the deputy high commissioner in India, and then from 2000 to 2003, the high commissioner in Pakistan.

Hillary Synnott: Thanks very much. Can I say, first of all, good afternoon ladies and gentlemen and it is tremendous to be back in India. My wife and I had three marvellous, and I have to say challenging, years here in mid-1990s and it is great to be back. I have already been back a few times since.

The first thing I want to say is that I am not an academic as you will have gathered and I have never quite understood the term discussant. This session today has not enlightened me much because clearly the word does not necessarily involve discussing what other people have said before one. So, I am going to take a rather different pack. What Mr Dasgupta did not reveal in those few kind words was that after my holiday in a very peaceful and quiet Pakistan between 2000 and 2003, I was sent to have a further rest in Basra, where I had civilian responsibilities for the four southern provinces of Iraq and that is what I would like to talk about now.

I think Mr Dasgupta has referred to the issue of nuclear proliferation as being a micro topic and you might indeed regard Iraq as a micro topic. But I think there is a tenuous linkage to the issue of terrorism partly because it has become part of the narrative, if you like—the narrative on the part of some in the Arab world, of the anti-Arabic, anti-Islamic tendencies of the West—and partly also because it has become part of the narrative of perhaps, shall I say, the less intellectually strong Westerners as being, 'Well these guys are bad guys. They are fighting us. They must be bad guys, and therefore, they must be terrorists.' So, I thought I will just talk a little bit about the current situation and the prognosis.

I think we will start from the fact that in 2003, when I first arrived in Iraq, there was still a lot of optimism around, unfounded as it proved, that things would get better quickly, that our liberation will turn into a liberation rather than what is currently characterized as an occupation. But I think the harsh reality impinged on our consciousness. I think with the first war in April 2004 and the first Fallujah incident—incident is an understatement—the destruction of the mosque in Samara, we saw a massive growth of sectarian violence, which added an additional dimension to the problems in Iraq. But I think when we are looking at the bad guys, it is worth just tightening them a bit, I mean not physically but intellectually.

When the West started off in Iraq, there were a lot of actually completely unfounded allegations, actually not a lot but one or two coming from very senior positions, that Al Qaeda were in Iraq and were causing mischief. They are now, but they were not then. Al Qaeda was/is part of the problem, but by no means the biggest part. I do not think it ever has been the biggest part. In the early days, it was what we call the former Saddam loyalists—Former Regime Loyalists (FRLs) in Army parlance, and no three words can be used without using the initials—but that developed into genuine insurgency, protests by particularly younger people against the presence of the West. That developed further into internecine rivalry between Iraqis because suddenly, paradoxically, the presence of Western forces in Iraq lifted the lid off Saddam's oppression and gave a whole lot of…to political forces who developed, and a lot of political forces and other forces were not necessarily benign. So, what we now see, particularly in the South where I was, is this very strong violence between what have been characterized as the pro-Iranian Bada Brigade and the pro-Muqtada ak-Sabah Jaish el-Mehdi, the Mehdi Army, and also in Basra other poltical forces, the Fadilah Party, who are now in government there, but crucially, just straight criminals.

What are they doing? They are not just fighting the West. They are actually grappling for a slice of power, which never existed before. Anybody who tried to get a slice of power under Saddam was simply assassinated. But now it is a free for all. That power is not only territorial power, where traditional Sheikhs are fighting, what were call Saddam's Sheikhs, Sheikhs whose power was boosted by Saddam, and for local territorial supremacy. We see this particularly in that province to the South-East, just next to Iran, where again people who do not think very hard say it is all Iranian mischief that there is so much violence there. It is true, there is probably Iranian mischief but there is also internecine tribal rivalry there. But the big goal in the South particularly is a cut of the oil revenues. There has been, of course, a smuggling system in Iraq that has developed in a very sophisticated manner throughout the 12-year period of sanctions long before the Western occupiers arrived. So, there are some big prizes to be had for the victors. The spoils are great.

Now, after these setbacks that I have described, we are seeing perhaps a resurgence of optimism and I would just like to qualify any such optimism. This optimism has a reason, as of last summer, and arising not just from the American surge of an additional 30,000 troops making a total of a 180,000 altogether, but I think more important than the physical numbers, a complete change of technique, of approach, guided by General Patreas's approach towards insurgency. That has had a very considerable effect on the ground. I think there is absolutely no doubt at all now that the level of violence particularly in the Central area, the Sunni tribal area, has dramatically reduced. This is partly due to the al-Anbar Phenomenon, which Nigel has referred to, where we see the local Sunnis have suddenly woken up to the fact that these Al Qaeda foreigners among them—they may be Sunni—are actually doing the local Sunnis no good at all. Not only that, they are doing them a lot of harm. So, the local Sunnis are turning against Al Qaeda and Al Qaeda is finding it far more difficult. I think the local Sunnis also have tweaked to the fact that Al Qaeda have been trying to foment sectarian violence and they succeeded in this to start with and I think that it is now reducing. There has been a…by General Patreas's forces, who are physically protecting enclaves of Sunnis and Shia from each other.

That is very good. As Nigel said, it is questionable as to whether the al-Anbar model is exportable to other provinces, certainly not in the South because there is no sectarian problem in the South, the people who are killing each other there are Shias. It is good but why do I have to qualify this optimism. What has been the objective since the start of the coalition forces in Iraq? It has essentially been to stabilize the place, provide a feeling of security to enable reconstruction and governance to flourish so that local Iraqis could feel that they have some future ahead of them and they have a stake in this future and if they see some green and pleasant pond ahead, they will want to walk along the path towards that. If they do not see any such good prospects, well why not join the insurgents and just protest because there is nothing better to do particularly for young men.

So, in a way the objective has been to create strategic space for the government with originally a lot of assistance from the coalition but now much less—to step in, improve justice and equity for the Iraqi people and to try and get the power stations mended, the water pumps going; the canals no longer fall in and the sewage does not fall into the canals. The difficulty is that the Iraqi governments at local and federal levels do not look as if they are either willing or able to fill that gap. So, if and when the American surge reduces again—that looks as if it may happen because there is a limit to the overstretch of the US forces just as there is clearly a limit of the overstretch of British forces—then what will happen? I do not know, of course. The future is turned. But let us just look at some possibilities, and none of them are really very good.

One of the possibilities, which has been mentioned for a long time and indeed is advocated, I think, by some US senators is that you simply cut your losses—well, I think this is a loss, of course—and allow, if not encourage, Iraq to break up into supposedly, to describe if I may use that word, three component parts. With Kurds in the North, war has been relatively peaceful ever since the two warring factions got together; the Shia in the South, and undoubtedly there is a Shia majority in the South; and in the middle, well, let us call it Sunni. So, let us say that there are three entities that emerge. There are two, to my mind, fundamental objections to that. One is that the picture is far too simplistic, and particularly in several levels (I agree getting too far) the Kurds are in a metastatal position—to use an engineering term, as I was an engineer. The Kurdish groups are together now. But they could easily fight each other again. The Sunni triangle is not a Sunni triangle. It is not a triangle and it is not Sunni. It includes massive numbers of Shias and the big cities are within themselves mixed up. So, if you were to talk about a Sunni entity, you would be talking about massive population movements. We have seen, historically, what happens in those circumstances.

The Shia south portrays a sort of a picture of amicable Shias that were all going to Friday prayers in the same mosque. That is not the case at all. They are very divided. I have already described that. But they are religiously divided also. There are those who are close to Najaf, there are those who are close to Iran and there are those who are close to Shia in Saudi Arabia. So, they are not an inherently stable group. That is the first reason. The second reason is because of the interests of the six neighbouring countries as to what would happen in those circumstances—an almost anarchic split in Iraq. I would say the three countries of the six, which are most significant in this regard—Iran, who have a great interest in what goes on in the areas near their border; Turkey, for obvious reasons in connection with the Kurds and we have seen how concerned the Turkish parliament is about all that; and the third is Saudi Arabia, who will be concerned about Sunnis. But in addition, you have Syria, who were not playing a very constructive role in the last few years; you have Jordan, who are the recipients of the greatest number of refugees I think and we are talking of millions; and Kuwait, who have been invaded once already, thank you very much.

So, the implications for the region could be very great and very dark. So, what are the alternatives to this? I am not offering a silver bullet to you, to use the American phrase. But there are theoretical alternatives, a more federal arrangement in which the desperate natures of those three broad regions are recognized more and, indeed, already in part of the existing Constitutional arrangements, the special mention of the Kurdish region has been acknowledged, which many people criticize for acknowledging too much and the Kurds say they have not been acknowledged enough. So, there is an acknowledgement of the special characteristics of the North, the Centre and the South. Those federal arrangements need not at all be on the basis of three parts but could be, say, seven parts where you have several of the southern provinces together but not all of them.

There is another alternative, which could be very dangerous and is a straight sectarian divide, a free for all really—Shia, Sunni, inter-shia, Sunni and Kurdish. Another possibility, which I think was a missed opportunity by the coalition in the early days of 2003, is to develop individual provinces more, give more power to individual provinces. I suppose, perhaps, a ridiculous power though but more like the German model. Of course, the coalition, and I was part of it, tried to battle all in all this and tried to sort it out, it created this Coalition Provisional Authority (CPA) in May 2003, a month after the 21-day war, which started on 20 March 2003. That CPA was [existed] (nobody really said how long it was going to be there)—this was not just my just guess, it was on the basis of the plan set out by Ambassador Bremer and others—for 2 or 3 years and a plan was set up accordingly, if you can call it a plan. But actually after 6 months of its existence, few months before the American elections, I have to say coincidentally, to wind up the CPA the decision was taken prematurely and it only lasted a year. So, the coalition tried and failed. The onus now is on the Iraqis themselves, on the Iraqi government. Can they do it? I do not know. It is a difficult task.

That leads me to my final point that in these circumstances, what on earth is the role of foreigners in Iraq? Should they not just get out and leave? Should not troops just leave? Are they not magnets for insurgents, bad guys, terrorists? In Britain, should not we stop our boys from getting killed for a deeply unpopular cause?

Tempting though it would be to support such a point of view, I think it would be quite wrong for two reasons. One essentially is an ethical reason, a moral reason. We quite frankly created a lot of this mess and I think we have a moral obligation to the Iraqi people to help them sort out the challenges that they face. That whole argument is not one which I think, unfortunately, will carry much weight in political circles but it ought to carry weight. The other is a practical one: I think that if a significant number of forces are not available to step in if things got really bad locally and if they were not available to continue to try and train Iraqi security foces, I really think that the risks which are described and which may yet come to pass will be upon us even quicker and with certainty while there still continues to be the possibility that things could get better. Before they would get better, I suspect that they may become worse. But they may still get better and I think we have an obligation to continue to try and make that happen.

Thanks very much.

Chairman: Thank you very much, Sir Hillary. I entirely agree with you about not being able to understand the word 'discussant'. I have been quite diddled about it myself.

We have now had four speakers who have made presentations on what I think is a subject that everyone is interested in. I would like to, as we did in the morning, open the floor for questions. If anyone would like to just make a point, need not necessarily ask a question, please do so. However, the rules of the game are that you have to identify yourself before you ask the question.

Nagma M. Mallik: What I am going to say is really in my personal capacity. It is not as a diplomat for India that I am speaking. My question is to you, Sir, Mr Raman. I came in a little late for your talk. But one thing that struck me was that you said that the causes of Jihadi terrorism are not contemporary, and that they are rooted in history, and that because of that the hearts and minds approach does not work. My own understanding is not quite congruent with that. I would say that what explains the picture to me, the pattern of recruits to terrorism in India today, it would appear to me that in fact the material conditions on the ground do make a huge difference as to whether terrorism gets started in the first place or not. I speak off the record. It is my understanding that the Kashmir issue and terrorism in Kashmir got a huge fillip because of the rigged elections of 1987 and that a lot of the terrorist leaders of today (some of them) were small time school teachers and people like that, who had stood for elections and who at the result felt that the vote was for them but they did not get elected. That just kind of manifested itself in anger. Someone was quoting to me the other day that Salahuddin was just a primary school teacher whose ambition was to become a Minister. He stood for the 1987 election and today he is a terrorist leader in Pakistan and a fanatic with a big beard and all the appurtenances of Islamic terrorism. My understanding is that it is due to material conditions on the ground that suicide terrorism of this type gets started, gets a fillip and then gets fresh recruits.

In Sri Lanka, the place where suicide terrorism was invented, it is the same thing—there is institutionalized discrimination against the Tamil minority in education, in employment, in government service, in the military, and so many things, through years and years of campaigning. I used to hear—when I lived there, when I was posted there—very emotional stories of people, who would tell me, 'My brother was just weaving on his motorcycle through the fields and the roads of Jaffna and the military convoy was passing and they just did not like the idea of a young Tamil boy looking so carefree and they just casually shot him dead.' Things like this and institutionalized discrimination and this kind of insensitivity and heartlessness on the part of the elements of government machinery—I think all these have provided the breeding ground for terrorism.

Similarly, there is the lack of terrorist recruits in the rest of India outside Kashmir, the whole mass of the other Muslims in the rest of India. The fact is that despite the poverty, despite the illiteracy, despite the unemployment, large numbers of Muslims from the rest of India, especially UP and Bihar where there is real poverty, have not become Jihadi terrorists. They are not members of the Al Qaeda. They are not terrorists. Some of them have joined the Mumbai underworld, yes. But they are not terrorists. That I think is because despite the Gujarat riots, despite everything else, the Constitution is fair. The legal framework in India is fair and not anti-minority. The minority feels that at the end of the day the systems are not stacked against them. They can go to law courts, schools, colleges, get into the government, get employment—these avenues are not closed to them. There is institutional fairness, there may be discrimination, maybe some people are bad and some situations may be bad, but institutionally, the legal framework is fine. I think that is why in India, as a whole there are no recruits to Jihadi terrorism.

Similarly, the causes in the Middle-East, I do not want to go into them—I am not an expert. But they are self-evident that there is a lack of democracy in all the Arab states with the US government supporting non-representative, repressive regimes. Israeli oppression of Palestinian people...approach along with policing, along with security, effective policing and effective security, there has to be a hearts and minds approach because I think that is what will eventually work. To my inexpert mind, the whole picture that I have just painted for you from Kashmir to Sri Lanka, I think that my interpretation is what fits these facts. If you have a different way of looking at these, I would welcome an explanation from you, Sir.

Thank you.

B. Raman: I was talking of international terrorism as represented by International Islamic Front. I was not talking of indigenous terrorism. If you have been hearing, following my policies, my interviews that I give, I always say we have to make a distinction between indigenous terrorism and international terrorism as represented by International Islamic Front. We must follow a hearts and minds approach in the case of indigenous terrorist groups. We must talk to Kashmiris. We must talk to Naxalites. Sri Lanka government must talk to the Liberation Tigers of Tamil Eelam (LTTE). Where indigenous terrorism broke up with a legitimate grievance—some are legitimate and some are not legitimate—we must talk to them and we must follow hearts and minds approaches. Where it is a question of international Jihadi terrorism, which has got a global dimension as represented by the International Islamic Front, they have got two global objectives. One is recreation of the Caliphate. Nobody can agree with it. Saudi Arabia does not want recreation of Caliphate; Turkey does not want a recreation of Caliphate; forget non-Muslim countries.

Secondly, we are getting mixed up...once upon a time formed part of the Islamic Umma. They want part of Spain to be given to the Islamic Umma.

Nagma M. Mallik: This is no emotive issue for Muslim youth, Sir, I am sorry. Recreation of the Islamic Caliphate and recreation of Umma, no young Muslim actually wants these things.

B. Raman: But these people are supporting them. You should not try to distort views. I make a distinction between indigenous terrorism and international terrorism. The entire topic was on international terrorism as represented by the International Islamic Front. I said we cannot negotiate with Al Qaeda; we cannot negotiate with Osama bin Laden and all. If there were legitimate grievances, we can negotiate. Where it is indigenous terrorism, I have been one of the strongest advocates of negotiations with all terrorist groups in India which do not have a global agenda.

Nigel Inkster: I want to exercise the prerogative to add a couple of comments on that if I may. I think my own perception in this is that the reasons why people come to international Jihadism are not uniform. They are very different depending on the circumstances of a particular case. Young men and women in Western Europe who come to the Jihadi schools tend to be driven by issues of identity; it is about recognition, assertiveness, nothing whatsoever really to do with religion. Very few of them have actually read the Quran and have more than a tenuous concept of what is involved. You look at somewhere like Saudi Arabia on the other hand, it does tend to be more religious arguments that drive people in the direction of Jihadism. In the case of occupied territories like Gaza, it is something else again. I think that for the use of hearts and minds, we need to be a little bit careful here. When you are talking about an indigenous terrorism where there is something to negotiate, that is one thing. I entirely agree with Mr Raman. Al Qaeda does not offer an agenda around which there can really be any negotiation. So, we are talking about using hearts and minds for a different purpose really, which is to try and create Islamic communities in which this particular ideology cannot take root, cannot flourish. That, it seems to me, is more of what we are looking at. It is about two things. One is encouraging the Islamic world to have a debate with itself about the sort of role Islam is to play in the modern world, and a separate debate which I think will arguably be applicable to India, certainly applicable in Western Europe, as to how an Islamic minority actually fits into and interacts with a modern secular state.

I think these are areas where I think intellectual debate and ideological debate have a place. I would characterize hearts and minds in that sense. You are not going to be able to win over people who have decided to engage in Jihadism, although I know that there are programmes in Saudi Arabia and Yemen to try and deradicalize young men who have gone down this route. It seems to me that what we probably ought to be talking about is trying to ensure they do not get to that stage in the first place.

Savitri Kunadi: I have heard with great interest the views which have been expressed by the various panellists. It is quite obvious that the subject of terrorism is really one of the most complex and politically divisive issue in the international arena. Earlier on during the day, we had also agreed that terrorism is one of the global issues for which we need a global response. I would like to hear from the panellists what kind of a global response we can really have. There have been various efforts made in the UN by the international community to form some kind of a global response. Until now the approach has been a piecemeal one whereby we have several agreements or conventions against different aspects of terrorism. A few years ago the UN adopted this counter-terrorism strategy, which I feel was a step forward. At the same time, it is not mandatory, it is not obligatory for states to pursue that or to follow that agreement that was reached. My question is about the kind of international global response that we can all come together and adopt, which would then make it much easier for us to counter this menace of terrorism.

Ishrat Aziz: I incidentally served for 23 years in Middle-East. So, that is an issue which has already interested me. I was a student of political science. What I found was that to develop a neat, coherent theory you need to ignore lots of very complicating things that come in. That is why, as a practising diplomat and who has not studied diplomacy in books, for me it was refreshing to hear Sir Hillary. The things made sense to me though they did not add up to a rigid theory. I would agree wholeheartedly with Nagma there. I have said it again and again that if you keep looking to theology, if you keep looking to scriptures and find out the causes of terrorism amongst any people, you are looking in the wrong place. Look to the circumstances around. I know the problem there. That is why I have stopped using the word 'causes' of terrorism. Once I use that, cause means that you are justifying it. If there is something as a cause, it has justification. So, now I say sources of terrorism.

Sources of terrorism are not in religion alone though religion has hijacked the news. I said this earlier also. Religion is used and it will be used. Ideologies will be used. Ideology was used in the past. Nationalism has been used for all kinds of things, which amounts to the same thing. I will give an example. How did nationalism, socialism, or fascism in Germany and Italy take people by storm? That is because they were frustrated and then here was an ideology that promised something very clear. When a terrorist comes with a very clear promise, a very clear-cut ideology, it has an appeal—whether that ideology has a basis in religion or nationalism or some other political ideology, that is not my point. My point is that unless you go deeper—and then the water becomes very muddy—you cannot just come up with a neat little theory. Unfortunately, neat little theories do not work in practice. If you look at the history, if you look at any leader who has really ruled a country, you will find he comes to power through an ideology and then jettisons it and does things which he thinks are right for his government and people or whatever.

Having said that, I have a question for Sir Hillary. I think he said something to the effect that Syria has not been very cooperative or helpful in Iraq. My question was, why would Syria be helpful? Is it a moral obligation or does it expect something in return, which it is not getting?

Thank you.

Ummu Salma Bava: My question is on linking up with perception, threat and responses to terrorism. I think even if you have an EU-agreed position on it, the challenge is how do you respond? Under the EU, we come under the second pillar, which means you put it down as a law and order policing function in that sense of justice, home affairs trying to respond to that. Then you have to take it beyond the domestic and you will have to get into the foreign policy position. You will have to come into the Common Foreign and Security Policy (CFSP) and we will not get any kind of unison over there. You go across the Atlantic and you have got domestic, internal responses over the Homeland Security Act, which aims at securing the citizens with a lot of intrusive elements.

I think you can see clearly that despite the so-called common threat and attacks faced, it is very difficult to get cohesion in terms of a response.

That also raises a fundamental question when you are looking at terrorism. The sub-regional, regional and the global is interconnected at some point because there are linkages happening all the way through. It acts as a lightning rod, which helps all these elements to come together. The critical question then to ask is, is it merely a law and order problem? Is that the route to take? Is it the military solution route? Is it about just, kind of, going to maybe the source issue? Or is it kind of something else or beyond? I do not think we get cohesiveness over there even if one is able to find a common minimum operational definition of what constitutes terrorism. I think that is where the perception of whether it is purely a law and order issue, of domestic realm, or are we going into international law and where—I think Navnita also correctly pointed out—we will enter into areas where we do not have a law, which deals with these so-called combatants who do not get covered under the so-called world definition, but are yet still waging a war, so to speak, comes into play. We enter into a grave realm. I am not talking about putting out because somebody would say, 'Oh! You are talking about rights for people or something.' I am talking about that. I am talking about how do you come to working definitions.

Nigam Prakash: I have served in the Middle-East. What we have to do about terrorism *per se* is to define what is included in terrorism. Is state terrorism also included? Because very often terrorism is used either during war openly declared or during peace times by state governments covertly and secretly. The incidents of terrorism in the last 40–50 years—or rather I would say after World War II—can be traced to historical facts and to certain grievances. The problem in the Middle-East is one major grievance, which I think has given birth to a large part of the terrorist activities and will continue to do so unless there is conversation, there is negotiation with some kind of an acceptable end in view. If one studies history of the last 50–60 years, one would find that this problem has been expanded by those involved in it in order to draw greater attention to it and get more and more people involved in it. So, I think we have to look at specific incidents, try to find out what is behind each and then try and solve the problem.

In India also, a large part of the terrorist activity against the country is based on historical reasons. People have certain grievances based on that historical experience, which the Government or the civil society in India, for instance, have not been able to solve. Then, these people who are involved, affected by it, try to find others to support them and, therefore, expand the circle of those who are affected by this particular problem. So, I think we have to look at it in two different ways. Specific issues have to be tackled. I would like to know whether the Institute has identified historical or factual problems of the last 60–70 years, which should be tackled at a global level by states concerned and by the civil society in general so that the sources or the problems are reduced and, therefore, terrorism which results from it is also reduced?

Chairman: Ladies and gentlemen, I think we are running out of time. If everyone agrees, we will just stop with one more question or statement from the audience. Ms Nagma Mallik wanted the floor to make a comment.

Nagma M. Mallik: The clarification that I wanted to make was, when I spoke of winning hearts and minds, Sir, I was not by any means saying that Government should try to win the hearts and minds of the Al Qaeda. I do not know if hardened terrorist groups like that have hearts and minds and if they have they can be won. I was talking about government policy vis-à-vis the ordinary people. The government has to just address its people, its constituency. Those are the hearts and minds that are there to be won, if there is any perceived sense of injustice that has to be removed. There is not even a need for an actual result. If the government is perceived as doing something with sincerity, automatically the sense of grievance, the sense of resentment inside people's hearts will be done away with. Of course, a major result is that there will be no new recruits to even international terrorist movements. But also, many people who are in the movement themselves, even though they have been there for years, I think even their own motivation will wane. That is my belief. That is what I meant by hearts and minds. I did not mean that governments should start negotiating with people like Al Qaeda and start trying to win their hearts and minds. That is not what I meant.

Finally, I would like to reiterate the point that Ambassador Ishrat Aziz made regarding the two points that Mr Raman, Sir, you made. I believe that these emotive and high sounding things like being part of the Muslim Umma, resurrecting the Great Muslim Caliphate, are just beautiful things to say. I do not think that people really mean any of these things. People understand perfectly well that we have come hundreds of years from that time and that these things are best left in the textbooks and the history books where they belong. I do not think anybody really wants to be part of any Muslim Umma. The Muslim community is too disparate, too different. The cultures, the languages, the food—they are all completely different. Nobody actually in any way thinks of being part of the Umma. The Muslim Caliphate is also just a very old thing. I really believe, I am not just making a point, that these issues are not emotive and passionate, not that the young Muslim people become passionate about.

Pradeep Sharma: I am not an expert on terrorism; my expertise is on Indian politics. I have a very small question to Sir Hillary on the argument of ethical obligation that he made. In fact, I find it very colonial because this has happened in India also and it sounds like a white man's burden of the 19th and the 20th century. If India would have done so in 1971, it would have occupied Bangladesh, or at present, it could just enter into Myanmar. I think this 'ethical' means that first you create the mess and then admit it is a mess; that also happened in 1947. This is just a small clarification on that point.

Chairman: Before I invite the four panellists to speak, I just thought I would draw attention to two points that Navnita made. Navnita said that no government in its fight against terrorism can adopt measures that break the law completely. If you continue to do this, it raises very fundamental issues and concerns. The second point that Navnita made is a concern that many developing countries have about the global fight against terrorism where we feel that Western countries appear to be adopting measures and policies that can become exceedingly discriminatory. The most critical one of these is, of course, the immigration policy. Genuine travellers find themselves treated very poorly. I am not entirely certain that this is a legitimate way to win hearts and minds or, for that matter, anything else, or indeed if it does help to create a global consensus against terrorism.

I take the opportunity now of inviting the four panellists. I may also mention that I have not been a very good Chairperson. We are running severely behind schedule. Perhaps, everyone could agree to speak for a maximum of about 5 minutes.

Thank you very much.

Hillary Synnott: I will try and be brief. There are quite a few issues raised and on each of them I have got an opinion, of course. But I would not cover each of them.

First of all, I would say that former ambassadors have in common with former generals, a tendency to say to themselves, 'Well, this is all very well, but what is the ground reality.' Ambassador, I think you and I could take tea together and agree on an enormous number of things. The point here is the politicization of religion or whatever the causes, the sources of terrorism. I have attended so many discussions on these, which are very interesting but often do not take one much further. I have in mind Amartya Sen's reference in his wonderful book, *The Argumentative Indian*, where he points out that at a time in the 16th century when Christians were burning each other at the stake, Emperor Akbar was preaching tolerance among religions. We do well to remember some of these lessons of history. They are not simple.

I think that leads one to Ambassador Savitri's question on the global response to terrorism. I think that is similarly very difficult because I do not think it is possible to devise a global response to terrorism myself. This is one of the reasons why I think the global war on terror (GWOT), apart from GWOT being an extremely unattractive word, is nonsense. Terrorism is a tactic; it is not an enemy. Having said that, there are some issues there, which one could draw on more and I draw a distinction, I draw a contrast. Consider the action in Afghanistan, which had the full backing of the UN, which is the coalition that goes much wider than NATO, although the attention is on NATO—it may be described in neo-colonial terms but it actually has the full backing of the UN. Contrast that to Iraq, which did not have the full backing of the UN. I know which I would prefer. Let us be clear, I mean the Afghanistan model and not the Iraqi model.

I think there is another aspect too, which is far more practical—which perhaps appeals more to generals and ambassadors and I think it addresses the Professor's point—and that is, how do we do this? Is this a law and order question? What is it? Is it a development question? The answer to that is that it is everything. Often we hear people say, 'Well, what is needed is a comprehensive approach.' What governments say is, 'What we are doing is we are taking a comprehensive approach.' They are not. What they are saying is, 'If only we can take a comprehensive approach!' There, I think, we have to look to our national institutions. I know, speaking for the UK, our national institutions are not designed for these sorts of missions, which really require a very strong agreement. Is it possible to imagine agreement between defence ministries, foreign ministries and development ministries? By definition, they disagree with each other. And yet, they got to have a coherent approach working together. You then have to expand that internationally. Look at the different national approaches in Afghanistan despite the fact that it is a UN-sponsored mission with these famous national caveats. It is very difficult, but I think that is what we should work towards.

On the fourth issue on Syria, I think, Ambassador, maybe you have responded to what I would call a typical British understatement. I said that, perhaps, Syrians were not being very helpful. What I meant was they were being distinctly unhelpful. I would turn the question round and say, 'Why should they be unhelpful? Why should they cause trouble?' That is what they were doing. I do not think they owe it to us to be helpful towards us. No! Who does? They pursue their national interest. But if Syria's behaviour in Iraq is part of their national interest, well, we have a clash of interests.

The final small point is about the white man's burden. These historical parallels with India, I think you drew it, is really very false. I would not try to defend the British colonial history in India. At that time, when, perhaps, my colleagues' grandfathers were in India, my grandparents were fighting the British in Ireland. So, do not label me with that one. But the big difference is that if there was colonial action, a parallel to be drawn in Iraq, it was about the original invasion, not about staying on. The Iraqi government wants us to stay there. The day they ask us to leave, we should leave. Of that I have no doubt. If we did not, that would be an infringement of their sovereignty. At the time of 1947, what was the

Indian government? It was British. It was the Indians who wanted us to leave. Again, I do not want to go into the history of India; you know it better than me. I would say that about Iraq, I think it is a false parallel. I think the fact is that the Government of Iraq wants the coalition to be there and as long as they do, I would continue to emphasize the moral obligation to stay.

Nigel Inkster: The disparity between theory and facts on the ground that has been raised is important. However, as a practical man whose job until recently was actually to catch terrorists rather than to talk about them, I tend to share that perception. It follows that I agree with Sir Hillary that in respect of global responses, I believe that we need to have realistic expectations about what can be achieved. A lot of the good work that is being achieved is happening bottom-up. It is where intelligence services, security services, governments themselves work out the need to develop more collaborative and cooperative structures. As I said earlier, probably the best thing that institutions like the UN can do is seek to create an enabling environment for this kind of activity to take place. To transpose the words of the Hippocratic Oath, the doctor's job is first not to do any harm. So, I think the international community's job should be not to get in the way.

Looking at the European dimension, yes, I agree that at the moment we have got a situation in which terrorism is a third pillar issue. It is by definition a national competency. There are again limits to what the EU can and, arguably, should do. In fact, within Europe there is actually a very close network of cooperation on terrorism, which I think encompasses all dimensions of the issue.

Is it a police and law and order issue? I agree with Sir Hillary. In the UK at the moment, for example, work is going on to develop a national security policy, which seeks to inject into all dimensions of all policy departments, the need to think about the terrorist dimension of the activities that they undertake. But making this happen in reality is going to be difficult and is going to take time. Another thing is that there are not any easy answers. Should it be a police problem? Fundamentally, I think, terrorism, yes. Sometimes, you may have to treat it as insurgency, as we are currently having to do in Afghanistan, and there, obviously, the military takes predominance. But I very much agree with what Mr Raman said earlier that terrorists need to be treated as criminals. My problem with GWOT apart from its disphonious sound—if that is the right term—is that it risks dignifying Al Qaeda with a status that they do not deserve. When we were dealing with the Irish Republican Army (IRA) in Ireland, I remember doing this in the 1970s, a time comes when negotiations are inescapable. We never gave ground on this point. We treated this as a criminal issue. I personally believe that that is the way to go.

One point that Sir Hillary did not pick up, a question was asked about whether the Institution is undertaking studies on some of these long running problems to see whether there is in fact any useful progress towards solutions by the international community. Yes, very much so. That is the very essence of our activity. All the time we are undertaking work in these areas. Frozen conflicts, different kinds of conflict situations where the resolution has been long in the coming. Yes, of course, we are doing work on that all the time. The information about what we have been doing already is available if anyone wants to consult our website.

The final point I would pick up is actually something that the Chairman mentioned about some Western nations in effect engaging in overkill and resorting to discriminatory policies. Speaking as someone who has regularly suffered ritual humiliation passing through US airports, I can entirely relate to that. Believe me, this discrimination is entirely colour-blind.

B. Raman: I will just make one or two points. There is a difference between counter-terrorism approach and the counter-terrorist approach. In counter-terrorism approach, you treat terrorism as a phenomenon which has got many factors—political, economic, social, law and order, security, etc. The counter-terrorist approach treats terrorism as a threat to national security or threat to international security. This has always been the policy of the Government of India if you see the way we have dealt with insurgency and terrorism by our own nationalism, when it was Al Umma in Tamil Nadu, or the Mizo National Front in Mizoram or the Khalistanis in Punjab or organizations like the Hizb-ul-Mujahideen, Jammu Kashmir Liberation Front (JKLF) in Kashmir, etc., we treat that as a phenomenon. We must keep in view the security approach, law and order approach, you cannot allow them to go around indiscriminately killing. At the same time, simultaneously, you try to find out what are the political reasons for that anger, what are the economic reasons for that anger and what are the social reasons for that anger. It is only because of this dual approach, taking them as a phenomenon where it is our nationals, our own people, that we succeeded in Mizoram, that we greatly succeeded in Nagaland—the Shillong Accord, now we are negotiating with the National Socialist Council of Nagaland (NSCN)—that we succeeded in Punjab. We have been trying to hold negotiations with the Naxalites or whoever it is. That is why we are urging the Government of Sri Lanka also to talk to the Tamils and try to find a political solution instead of trying to suppress them by force. That is where your own nationals are concerned.

But where foreign terrorists operate, we have got two kinds of foreign terrorism. Pakistanis are coming here, like Lashkar-e-Tayyaba—what are its objectives? In its agenda, initially, it said Kashmir. It said that Hyderabad should have gone to Pakistan at the time of Partition. That is why it has set up a branch in Hyderabad. It said Junagarh should have gone to Pakistan at the time of Partition. There were a number of things that it had mentioned about the Hindus and all. It is not a question of hearts and minds approach to foreigners who have got no locus standi here. They come into our territory and operate, whether it is Lashkar-e-Tayyaba, Harkat-ul-Mujahideen or HuJI; whoever it is, we have to treat them as a threat to national security and we have to take a security approach to them. There is no question of taking them as a phenomenon. If Bangladeshi migrants start agitating tomorrow, will you treat it as a phenomenon and not take any action against them? No, you cannot do that.

The Afghanis and foreigners, the Arabs and all—whether they come here or they go to the US, or they go to Europe, etc.—they have to be treated as a threat to national security. They cannot be treated in any other way. Yes, if a British Muslim has got a grievance, the French Muslim lady has got a grievance because she has been told that she cannot cover her head to go to public school, yes, we have to talk to them and convince them as to why the government has to take action. If there is legitimacy in that, we should talk. But somebody from some other country goes there with explosives and if somebody goes to Denmark and blows up an atomic bomb there, if somebody wants to destroy your maritime trade, etc., you cannot just talk about hearts and minds approach and say let us talk with them over a cup of tea. We have to treat them as threats to national security, as threats to global security.

The other point, which was made was that nobody supports it, they just use it for rhetoric. If nobody supports it, why so much terrorism? Terrorism is not coming down. So, many people are going and joining terrorists. Nobody supports all this. Everybody treats Osama bin Laden as a mad cap to talk of Islamic Caliphate and all. Everybody treats International Islamic Front as an irrational organization. Why are more and more people joining this? There may be other reasons but they all go back and agree with the pan-Islamic objectives of that organization. So, it is a highly romanticized view to say that nobody is joining, it is only rhetoric, etc. Anybody who has responsibility for national security cannot afford to take a romanticized view.

The last question the Ambassador has asked is on global response. I have made a point. I have said one of the problems which we face today is because of over-militarization of counter-terrorism ever since the Americans came into the picture. I will say that again and again; they used the air force, they used artillery and all. We have been facing this problem of insurgency and terrorism for long. Once we used the air force in Mizoram in 1960s. Thereafter, we never used the air force. We never used heavy artillery whatever be the provocation, whatever be the casualties, etc. So, our approach has been to try and use only small arms. But unfortunately, what has happened is the Americans are fighting their war against terrorism outside the US and not in their own territory. They are not fighting the war against their own nationals, they are fighting against foreigners. So, their doctrine is to use the maximum available force in order to defeat the terrorists, be it in Iraq, be it in Afghanistan. They want Musharraf to create the tribal areas. So, this over-militarization, over-Americanization of counter-terrorism methods are leading to additional anger in people who were not angry earlier. That is why I keep on saying that in India till today Muslims have not joined the Al Qaeda. In fact, they have kept away from Al Qaeda because of our neutral policy towards the US. Tomorrow, if they find that we are associating ourselves with the American counter-terrorism approach with military and all, then they will be drawn into terrorism. So, there is no question of global response to international terrorism because we have to take into account our own country. We have the second largest Muslim population here. We have to take into account their sensitivities, their feelings, etc. So apart from sharing of intelligence and all I do not think in dealing with this we can have a convergence of views as to what should be the global response. One has to decide depending on the circumstances of our populations.

Navnita Chadha Behera: I think my task has been made really easy by all the three preceding speakers. Most of the points are covered. There are just two very quick and short points.

At the global level, I agree with what Mr Raman ended that there has to be modest expectations as to what we can achieve at a global level in terms of achieving a consensus. As Professor Bava was pointing out, even on simple things like working definitions, we still have a long way to go. SAARC had arrived, for example, at a definition but it is practically defunct. Even on theoretical propositions of just evolving working definitions to issues of strategy, to issues of developing ethical measures and counter-terrorism strategies, there has to be a global sort of a consensus. But I think the battle really lies in the domestic domains. There it has to be very variant, responding to the ground situation.

The second element, it has really been joined in full, is whether the hearts and mind approach is relevant and who should it be addressed to. I just want to make one point, in fact. You said it very rightly that you meant probably hearts and mind approach to responding to public, which is caught in the crossfire, so to say. I would go even one step further, and as Mr Raman pointed out, say that you also have to address the political grievances of the community that is resorting to violence. I think that that is where the Indian state has achieved maximum score in terms of developing a resilience of power-sharing mechanisms.

Ultimately, the parameters are that, yes, if you were to be able to give up the violence, we would negotiate with you and, ultimately, we share power. That is how the Mizo problem and the Punjab problems were resolved. Many accords have shown that the state has developed resilience in being able to share power with communities that are posing a challenge to it. But at the same time, there is merit in the argument that there are many groups that are actually not amenable to negotiations because they themselves do not want negotiated solutions. Their objectives are beyond what you can assume to be negotiated spaces. With those groups, you do not really have an alternative but to take a hard security approach and work towards elimination. I think it is a combination of two. The effectiveness of the approach would depend on whether it is an appropriate mix. That is what I was mentioning. It is the right mix of the political and security measures which decides whether the strategy is working or not working.

Chairman: May I take the opportunity of thanking all the speakers and members of the audience. It has been a very stimulating discussion. The good part now is that I get an opportunity to invite you for tea and cakes. The bad part is that the break is only for 15 minutes.

Thank you.

[*Tea Break*]

Session III

Energy Security

Speakers

CHAIRMAN: Amit Dasgupta, Joint Secretary, Public Diplomacy MEA
COL. (RETD.) JOHN GILL: IISS Lead Speaker
DR SHEBONTI RAY DADWAL: Indian Lead Speaker
RAHUL ROY-CHAUDHURY: IISS Discussant
DR J. NANDA KUMAR: Indian Discussant
UMMU SALMA BAVA: Professor of European Studies, Jawaharlal Nehru University, New Delhi

Chairman: We now come to the third and final session of this Foreign Policy Dialogue. I must confess that I thoroughly enjoyed the previous session. Unfortunately, you will see a rapid transformation in the personality of the Chairperson, who will shed his benign nature and put severe curbs on speaking time. We will close this session exactly at 5.25 p.m. The earlier decision was that the speakers would get 15 and 10 minutes, respectively. That stands reduced to 10 and 5 minutes, respectively. I am sure no one is going to like this decision, but I am afraid we have to shut shop exactly at 5.25 p.m. Please understand our compulsions.

Thank you very much.

John Gill: Good afternoon, I thank our friends in the MEA and my colleagues in IISS for the opportunity to talk to you this afternoon on this topic of energy security, a critical topic, indeed, for all of us. I have to start with a disclaimer that although I am currently affiliated with IISS, I am in my other life in my reincarnation as a US Government employee. So, I have to stress that my remarks represent my own personal opinions only and not any sort of official US Government representation. So, I would appreciate if you could treat those as not for attribution.

A few of the background points that Rahul and our MEA colleagues asked me to address. Energy security is a fashionable topic but that does not make it any less important. It is vast, it is complex and as others have noted, it is elusive. I think we can take 'the availability of energy at all times, in various forms, in sufficient quantities and at affordable prices' as a working definition for what energy security might be. So, this is a topic on which we could have spent all day or all week in looking at various aspects.

I would hasten to add that I am not an expert on energy per se. My specialty lies in security and foreign affairs. So, I will address some of the external dimensions in this quick time period and not things such as internal reforms and market structures. Geographically, I will focus on the Indian Ocean region that is reaching from the Gulf of East Asia; in other words, the context is to be India's area of immediate strategic interest, and I will look out to about 2030. I hope, thereby, I will generate some interesting discussion.

With that as a background, let me start with a few structural assumptions as we think about energy security in the Indian Ocean up to about 2030. First is the importance of Asia. We can welcome ourselves into what people are calling the Asian century and whether this represents a rise or a return of Asia to prominence on the world stage, it is clear that Asia as a region is going to be assuming a position of global significance economically, politically and militarily in a manner that it has not occupied since the 1500s. This shift is driven largely by the rise of China and India. For our purposes, this has big ramifications for energy. Indeed, the *World Energy Outlook* published in 2007 notes that India and China are transforming the global energy system by the dint of their sheer size and their growing weight in the international fossil fuel trade.

Second assumption is that world energy demands will grow by nearly 2 per cent per annum, that is, 55 per cent between 2005 and 2030, and is likely to continue to be dominated by fossil fuels as key source for energy after this 2030 time window. Although we can hope for a rapid development of alternative energy sources to revise this outlook, in the absence of any major change in fossil fuels, it works out for

84 per cent of increase in demand. Everyone knows oil will remain the single largest fuel but to some surprise, actually coal will grow as a component with India's coal use tripling by around 2030. So, this poses, of course, enormous environmental concerns. My esteemed colleague will address those in some more detail.

Third, Asia and India, India especially, will have to contend with serious vulnerabilities of the security of energy supplies. These fall into what I can consider two categories—one is lack of diversity, in short, you cannot replace the Persian Gulf or Arabian Gulf as you prefer. Asia cannot live by something of two-thirds of its oil from the Gulf but that is likely to grow to three-quarters by 2030. In India's case, the proportion comes to approximately two-thirds today and mostly from Saudi Arabia and that too will increase over time. So, the Gulf will remain the single most important source for oil and natural gas for the foreseeable future. Indeed, Organization of the Petroleum Exporting Countries' (OPEC's) share of output alone is projected to grow from 42 per cent today to 52 per cent in the next two decades.

The other category is reliance on sea movement. Just as today, the oceans will remain the likely key avenue for energy transport for the next two decades. It is possible that some of the pipelines that are under consideration will be viable by that time period but it is not at all clear that they will be. Even if one or more of them do become viable, there is clearly going to be a significant gap before that takes place. This continues then to draw our attention to the famous sea lines of communication or infamous locks in our world, particularly through the Indian Ocean. So, these broad projections leave me with some interesting conclusions.

First, that we will see greater roles and responsibilities for both China and India in influencing the character and direction of the global energy system and the associated issues. With energy security as a prime example of interdependence among states, this suggests the need for closer coordination with both of these powers on these key issues and also between India and China. Second, this centrality of fossil fuels has several consequences. Our focus is on the continued importance of the Gulf and the Indian Ocean in this regard but, of course, there is a key environmental impact because the unchecked fossil fuel use, if it continues as it is practised today, is going to be detrimental for all of us.

Third, the security of fossil fuel producing states in the sea lines of transportation that connect them to consumers will be increasingly important as national security priorities. This is especially true for the Indian Ocean and the key Straits on the eastern and western ends, the Straits of Malacca and Hormuz and, above all, Mandep—somewhat less so. But, in my opinion this is not solely a national concern because these are truly global public goods—these transit routes. So their security is central to national interests of friends and allies as well as our own countries. In the US's case, the security in the Indian Ocean is key for the successful rise of India as a major power that the US has stated it wants to support since that is reliant on economic progress, as the Foreign Secretary reminded us this morning, but beyond that, beyond the Straits of Malacca to the East, the economic well-being of America's treaty allies such as Japan, the Philippines, Thailand, and many others is also a key US national security interest.

Moreover, in my own personal opinion, the continued economic viability of China as a stable progressive country integrating peacefully into the international system is in everyone's interest and I would suggest that these arguments apply to India as well as they do to Washington…couple of security points in particular. First, the threats. We will have to contend with what I think we can characterize as two basic types of threats. The first is direct physical threats from terrorism. Mr Raman highlighted a lot of these very eloquently. That includes attacks on processing facilities, in both the producing and the consuming countries such as those with refineries, as well as terrorist attacks on shipping, either to create an economic impact or to produce mass casualties. Here look at natural gas or liquefied natural gas (LNG) shipments—although these are heavily safeguarded, it would be especially worrying if terrorists are able to overcome the safeguards with one of those ships in a port.

Second, the threats to socio-political stability of the producing states, especially in the Gulf. Most of these countries face serious internal issues that could result in domestic unrest affecting their ability to be reliable suppliers. There is also the possibility, of course, that nation states might act to inhibit energy shipments. Most often mentioned in this regard is Iran in the Strait of Hormuz, but for me, I see these more or less as unique case-by-case, and handled on a case-by-case basis. Higher sea on the other hand, which one often also hears about is, I think, less of a threat as I understand these things, my technical perspective of oil or natural gas shipments, because, after all, if you take the super tank, what will you do with it? However, piracy or the spread of piracy could hamper other kinds of commerce and contribute to a general sense of instability and uncertainty with possible economic and security costs.

In response, what do we think about these? Or the significant challenges we face in countering these threats. As regards the terrorist threats to shipping and facilities, there are unilateral and multilateral angles to consider. At the individual country level, the measures to be taken by individual countries to secure storage and processing facilities on their own soil are fairly obvious. However, each country has to develop strategies for securing shipping as well. Again, at the level of the individual state, that requires a lot of difficult decisions on things such as size and structure of military forces, resourcing of those forces and balancing energy security requirements with other national security priorities. The goal would generally be to emphasize robust and flexible maritime forces, but the requirements likely would include some sort of minimal expeditionary capability that would involve all military services.

We shift to the international side. There is no single nation that can cope with these threats to shipping on its own. So, part of any energy security strategy will have to be international cooperation. In the first instance, this means intelligence sharing on possible threats and preventive measures. Again, Mr Raman addressed this very comprehensively as did Mr Inkster while talking about terrorism. It equally applies to terrorism with regard to energy security. But it also means close and continued interaction with foreign militaries. India, the US and others have made some very significant strides in this arena in recent years and we can look forward to future progress. This is crucially important because establishing a deep level of understanding and familiarity in advance of crisis will be a key to responding effectively if an emergency arises. You cannot expect to turn to General Malhotra and say that tomorrow we have to go and react to a situation if I have not talked to him in advance. So, it is useful to know one another before we get involved in an emergency, so that we have an understanding that is built up on years of familiarity.

The other challenge beyond these military dimensions of energy security is in the state's stability among producing countries. Clearly, this is a critical national interest among consumers such as the US and India and others, but it is one where our ability to promote stability is limited. Indeed it is where ends, ways and means are very murky and very controversial. So, let me close, here, with one observation and several questions. The observation is that if this general construct that I have laid out very briefly here is accurate, then close and continuous international coordination is imperative in ensuring continued security of energy supplies. Therefore, we need to consider what international mechanisms might be best suited for this sort of an imperative. Some forums already exist, but we need to think about whether those are adequate and if not how they could be modified or supplemented to support the security of these global public goods.

In terms of questions, the requirements of energy security also pose a number of daunting questions, many of them relating to the structure and realities of international cooperation that I just suggested as part of an answer. In the first place, what role would the emerging Asian energy giants, India and China, be willing and capable of playing in some sort of a scheme of global energy security focused, in our case, on the Indian Ocean? This is related to how the two countries will manage the collaborative and competitive aspects of their energy-relationship. We can refer to the comments that Governor S.K. Singh made earlier about his thoughts on the role India and China might play.

Second, how will we mitigate the frictions over very difficult issues where India and the US, and/or other countries might disagree such as Iran, Myanmar or Burma? Third, at a more practical level, all of the interested countries need to consider what military forces are best suited to securing energy transit lines through the Indian Ocean in these important straits and how we might complement or supplement one another to foster common security—the key threat is non-state actors who are enemies to all of us—where states still have competing national interests.

Finally, what can outsiders do to foster domestic stability in producing states, especially in the Gulf? We have seen lots of discussion on democracy, on political reform from all sides over the past several years, but these remain as controversial as they are important. So, there are no easy answers there by any means. With that, let me thank you for your attention and return the microphone to our chairperson.

Shebonti Ray Dadwal: Thank you to MEA for giving me this opportunity to address such an august body. Jack has already made my task much easier and the clock is running. So, I will not repeat what he has said except that I would like to say that as far as some of the reasons why, today, the energy security has come back to the forefront of national security issues are concerned, I agree with almost all of them but with some of them, of course, I do not agree. However, the major challenge, to my mind, with regard to the energy market is a manner in which some of the actors are responding to the changes that are coming about so rapidly in the energy market. I think the most dangerous aspect of this is that some countries are adopting a very nationalist approach to energy security and even going as far as being ready to use force, military or economy, to protect their energy interests. There is even talk of some of energy consumers forming a cartel to counter the producers' cartel. Others, of course, have shown more understanding of the need for collective institutional measures. I am, of course, in favour of the latter approach that these problems need to be sorted out by the international community in a collective manner and the divisive trends that are emerging in the global market have to be replaced by cooperation.

I also believe that energy security is an issue that is of common concern for both producers and consumers alike and has to be, therefore, addressed jointly. After all, availability of energy is a matter of universal concern and with producing countries now using more energy than ever before for their own economic development, the pool of energy resources is going to diminish as they are non-renewable. At least oil and gas and hydrocarbons are, unless they are replenished timely. Therefore, I would say that the answer is to increase the pool of energy resources or to find alternative energy resources. This brings me to the issue that I would really like to discuss, which I think is a more contentious issue today with regard to energy security and that is the linkage between energy and environmental degradation.

As in the case of oil and gas where industrializing countries like India and China are being blamed, if I may use that word, for shrinking the pool of energy resources in the world because of their rapid and high consumption, the industrializing countries also are being held responsible for contributing to the increasing global carbon emissions due to their growing energy consumption. I would like to say that despite the increase in our energy intake, we are still far behind the developed countries in terms of our energy consumption. Our per capita consumption of energy in India remains one of the lowest in the developing world.

Having said that, the state of the global environment is a matter of concern for all—the developed countries as well as the developing countries. The adverse effects of environmental degradation will be faced more by the developing countries. A recent report by the New American Society and Center for Strategic and International Studies, in fact, warns of the environmental and national security implications of a rise in global temperature under three scenarios. All three scenarios paint a very catastrophic picture. Of course, there is also evidence of the fact that the GHG emissions have increased by 20 per cent since 1997 and this would increase further and faster and could grow by something like 60 per cent by 2030. India has been at the forefront of the environmental protection movement. But we insist that environmental concerns need to be addressed collectively in a cooperative and just manner. I believe that just because India has embarked on a path of development where access to primary energy resources is an imperative need if we have to meet our development goals and alleviate poverty, to now say that we must restrict our use of energy, as we are one of the largest polluters in the world, is not fair to my mind.

As the Foreign Secretary said this morning, our economic growth has averaged about 6 per cent over the last 20 years, and over the last 3 years, it has been averaging about 9 per cent. If we have to increase or even sustain this level of growth, we need access to adequate energy supplies. Even today, we have got a section of some 600 million people in the country where energy consumption has been restricted to about 4 per cent since 2004 despite our GDP growth levels being about 9 per cent. Our energy intensity levels too have come down. Our integrated energy policy that was brought out last year has laid stress on increasing the share of renewable and clean energy resources in our energy mix.

Because of paucity of time, I will not really recount the number of steps that the Indian Government has taken. If anyone is interested, I can talk about the various steps. I do not think a lot of people are aware of exactly how much the Government is really doing to address the environmental concerns.

Having said all that, I would say that though climate change is a global concern we must not leave the dispersal and proliferation of cleaner energy technologies to market forces and commercial interests alone as these are not going to be affordable or accessible to those countries and to people who need energy the most and are the fastest growing. There must be commitment for the common development and sharing of cleaner technology if we are to pursue the twin goals of economic development and environment protection. As I have said before, just as access to conventional energy sources in the long-term requires international cooperation and a collective approach, the same is true for facilitating the transition to a non-fossil fuel or cleaner energy sources. For this, the developed countries should and must contribute to sustainable development by providing adequate funding and support for the development of cleaner technologies in the public domain for adoption by the developing countries.

In conclusion, therefore, I would like to say that the answer to everybody's energy security problems is moving away from a fossil-fuel based economy to cleaner, green fuels-based economy. This may not be possible, of course, in the short- or even in the medium-term. But the important thing is to act now. The way we are going about things at Bali and even in other areas of energy, just fighting and going and becoming more competitive, is just going to exacerbate an already tenuous situation. Instead of just talking, I think it is about time we started implementing some of the very good ideas that have already been put forth in many quarters.

Thank you very much.

Rahul Roy-Chaudhury: Thank you Chairman, I have 5 minutes and a fairly basic but fundamental point to make that India's foreign and security policy will increasingly be driven by a search for energy security. What I would like to do in the next few minutes is to focus on two key issues—India's current and emerging energy diplomacy and India's emerging competition with China on energy resources.

In terms of India's current and emerging energy diplomacy, the key of course is the Gulf region where over two-thirds of oil imports are sourced from. Four days ago, India's National Security Advisor to the Prime Minister, Mr M.K. Narayanan, was at the IISS Manama Dialogue, where he spoke on energy and regional stability. He pointed out a key factor in terms of external partnerships and the importance of external partnerships—that India sees external partnerships as playing a vital role in bridging its energy deficiencies—and noted the synergy between India and the Gulf in this respect. He added that the importance of the Gulf in India's strategic calculus had only grown with the lapse of time. I think, we will continue to see this.

At the same time, there is another emerging area of importance on energy for India and that is Africa. In the past few years, Indian petroleum companies have begun to focus on Africa. Africa today holds about 10 per cent of global energy resources. India's largest source of oil from Africa is Nigeria, a traditional source of oil. The Indian Prime Minister was in Nigeria a few weeks ago, a visit that took place after several decades. More important, I think, is what we may be seeing in the next 2–3 years, an Indian focus on accessing and investing in Africa for energy resources. Last month, India hosted for the first time, the first India–Africa Hydrocarbon Conference, and the first India–Africa Forum Summit takes place next April in New Delhi. So, I think, a key area of future energy diplomacy for India will be Africa.

Secondly, in terms of competition with China, we have begun to talk about India as a rising great power. China has also risen in many ways in terms of its energy requirements and its demand. Today, China is dependent on oil imports for 40 per cent of its total energy demand. But the key aspect is that 70 per cent of this 40 per cent imported demand is sourced from the Persian Gulf and Africa, which means that we are looking at an increasing number of ships, Chinese flagged ships, traversing the Indian Ocean. What we have seen in the recent past is a mix of cooperation and competition between India and China on energy resources. Both countries have competed with each other in various parts of the world. In Angola, in Kazakhstan and in Myanmar, we see that continuing. But they have also cooperated with each other in terms of joint ventures in gas fields in Iran, in the Greater Nile Oil Project, etc. There is a sense clearly that as both countries require greater energy resources, the competitive element will be more important and will actually be more dominant than the cooperative element.

To conclude, I think the key really is the management of this energy security relationship between India and China. An important challenge for India's foreign and security policy will be to balance the competitive and cooperative relationship with China on energy resources on mutually beneficial terms. And, such a relationship will become more publicity prone as India follows China in investing in conflict-torn areas of the world. Both countries start investing in conflict-torn areas of the world. And, if you look at the Indian Ocean, it will be critical in many ways to build Sino–Indian naval confidence in this area through maritime cooperation and diplomacy. This would encompass measures such as the establishment of an official Sino–Indian bilateral dialogue on maritime security and cooperation on issues such as search and rescue, maritime safety and security, and marine environment protection and preservation. A possible way ahead for future Sino–Indian cooperation on maritime security issues—based on the inevitably growing Chinese demand for energy resources travelling the Indian Ocean—has already been published in a recent IISS Strategic Comments. But I would just like to emphasize what Governor S.K. Singh said to us earlier today. Namely, that there is clearly a need for cooperation between India and China in the Indian Ocean. My sense is that this is a key trend that may well be operationalized in the future.

J. Nanda Kumar: Thank you Chairman. Panellists have already covered many of the important points. But let me start with this.

Being an energy researcher, I have always had a question, that is, is India's energy consumption posing a kind of energy security challenge or energy security threat to the developed world? Often, after reading many of the write-ups from Western scholars, I felt so. I believe that the international community must work together. Energy security is actually not an individual responsibility. The international community must work together towards meeting the energy needs of the people in the world. As Shebonti mentioned, we actually must work towards ending the blame game so that we can actually work for a positive future.

By saying this, let me come to the Indian scenario. Here in India, enhancing energy security is actually guided by three main pillars, which are accessibility, availability and affordability of energy resources. Accessibility is certainly about how well are you actually connected with the energy supply countries and supply regimes; availability is about availability of different types of fuels, whether coal, oil, gas or other fuels; and affordability certainly talks about the price which we have been seeing in the global scenario. Prices actually have gone up to a level that developing countries are finding difficult to adjust to. These are the areas where we would like the international community to work together instead of carrying on with blame game.

Even today morning, Patrick has mentioned that the rise of Asia is actually seen as an opportunity. If you see it to be so, what, actually, we should do is that we should work towards that. We should keep supplying this region or we should be creating an environment for these countries to access energy resources so that their economy can move on, they can address many of the serious challenges within the country like poverty, unemployment, etc.

What can we do for this? I would obviously see that for India, a certain level of energy transition is required. Energy transition means not shifting completely away from the traditional fuels, but at the same time, making an effort to build up energy sources so that our rural as well as urban areas can gain access to energy. At present, most of the rural population is still dependent on biomass burning, which certainly contributes to environmental hazards. So what we need to do is develop our renewable sources. Obviously, India is one country, probably the only country, with a Ministry specifically focusing on renewable energy.

I believe we have been doing quite positive work towards this. But what we require at this moment is the cooperation from the international community, basically the developed community, which have developed technologies. I feel it is important for these countries to focus on not leaving these developed technologies for clean energy development to the market forces. That is because what we saw in the case of global price rise is basically because of the market forces. It is because we saw that energy is one commodity the big corporates are using for listing themselves in the stock exchanges. They want their commodity price to go up. When the commodity price goes up, their share obviously goes up. So, they are making money. I believe every producing company is making almost 20 per cent profit from each barrel of oil. So, in that kind of a scenario, any oil company would like their oil to come up only when the price is high. Otherwise, they would not even dig the oil.

What I would like to emphasize here is that there should be a kind of consensus among all the countries basically for supporting developing countries. We should address the needs of developing countries in developing renewable and new technologies.

With that, I would stop here. Thank you.

Chairman: Thank you very much. At the outset, I would like to apologize. I do not want to give the wrong impression to the audience that we are rushing through this section because we do not consider the subject or the topic to be important. But our guests from IISS have an appointment, which has forced us to stick to time constraints.

Ummu Salma Bava: The issue of 'cooperation' at the international level sounds simplistic because it is a public good to offer 'philanthropy'. There is no philanthropy in international politics first of all, let me put it out there. When there is no philanthropy, we are not getting into public choice debate, of a public good and, therefore, there is collective interest. Everybody understands that there is collective interest but what we actually have is a cartelization of production. As long as that exists, you will continue to have this kind of a problem. So, this expectation of a collective cooperation at the global level is something I find very difficult to see actually getting transformed into action.

The second thing was on technology availability. This has constantly been brought up. It is also part of the discussions, even at Bali, on technology. The point is, you have governments talking to each other. Governments are not manufacturing technology. Technology is being made by firms. They are not part of the negotiating instruments. So, when you are looking at who is going to provide the technology, it is going to be a firm. Is that firm willing to believe in, again I come back to it, philanthropy to just give this technology away to somebody else? It is not going to happen. That is where the interface of the state comes in. Will the governments then be willing, will the developed world be willing, to create the so-called technology pool to access and make it happen? There are no free lunches. What is the price tag attached to that? Would it be a low economic growth for the developing countries as reciprocity to receive some of this technology aid? I think that is a critical question to answer.

A third thing on clean fuel. Anything which is sourced from the earth is going to have a by-product; anything which is going to be converted, other than the solar cell—I do not know what kind of a negative by-product is there other than photovoltaic cells; we have too many of that all around. This whole notion of shifting to biofuels again is very dangerous. Farmers in many parts are going to be shifting from food security position and we are going to convert very scarce water resources, and our food cycle is going to be affected, to shift to just using a biofuel because that is seen as a green alternative. You have to ask at what price is this green alternative going to be produced? At the cost of food security? At the cost of farmers being again pushed out into a cycle of not being part of it and where we have a whole lot of other systems dependent on agriculture? So, I think these are some things which would need to be answered.

Most importantly, the cooperation at the international level, just because it is a collective thing. We all realize that. We all know that war is bad. Do we all cooperate? We would hardly need any convention or any war (If we understood this).

Chairman: What I would request is that John and Shebonti, if they would like to, make brief closing statements after which we would consider this session closed.

Shebonti Ray Dadwal: About collective cooperation and collective responsibility, the oil market works this way. Of course, there are some countries, the problem is that some of the oil, producing countries have only that one resource to sell and that is oil, and their revenues and their economic growth depends on that. But when you push that too far as they are doing right now, I mean prices are phenomenally high right now when there is really no need for them to be that high. At no time in history has it happened that this movement towards alternative energy resources has taken up... It is not just the environmental aspect of it; the price factor also plays at this point of time. So, it really suits the producers as well as the consumers if the price is right, not too high not too low. That brings you back to the definition of energy security. Energy security for whom—consumers or producers? It really works both ways if there is a kind of a consensus on how much to produce, setting the price right, etc.

Second, about this whole spread of nationalism, in a lot of the energy-producing countries where a lot of their sectors were nationalized, that restricted inflow of technology as well as investment—which is really one of the reasons why the supply pool is drying up today because they are not opening up their energy sectors to foreign investment—if that was allowed, if they would open up their energy sectors to foreign investment and allow the kind of state-of-the-art technology needed to enhance production, that could resolve the problem and enlarge the energy pool. That is the kind of thing I meant by saying that we need cooperation and understanding amongst producers and consumers, not the kind of exploitation that was resorted to by some of the oil companies earlier, before the nationalization process set in, but to do it in a collective manner where everybody would benefit and the resource pool would be increased.

The second point I would like to take up is about the biofuels. Yes, you are right, this whole debate about biofuels—whether to get energy security, you have to compromise on food security. I can speak for India, for instance. This whole thing about our ethanol and our Jatropha plantations, etc., we have a lot of arid land, which is really not used for any other kind of food grain cultivation, which is really being put to Jatropha production. Even in our ethanol production, I think we are using the waste from sugar crops, really, to make ethanol because we are acquiring the technology from Brazil where we are not really using sugarcane as such for production of ethanol but rather, the waste products. So, that is not really going to interfere with the production of biofuels. That is as far as India is concerned. But yes, I am completely in agreement with you that this is a dilemma in many countries and the more we move towards biofuels as an alternative to oil and to petrol and diesel, it is going to have repercussions on food security.

Chairman: Thank you very much. I must say that I am grateful to all the speakers and the members of the audience for their understanding. It remains for me—and I know the IISS joins me in this—to thank all the members of the audience for their active participation in the deliberations. These have been extremely stimulating discussions.

I know the members of the audience join me in expressing our appreciation to the delegation from IISS. We enjoyed this full day's session. I am authorized to say that, in principle, we are very happy to take this dialogue process forward on issues of mutual interest and concern.

Thank you very much. Good night.

Appendices

Appendix 1

MEA–IISS Foreign Policy Dialogue
New Delhi, 13 December 2007

PROGRAMME

1000 hrs.   : Inauguration of the Dialogue by Foreign Secretary

1030 hrs.   : Tea Break

1100 hrs.   : Session I–*The Strategic Shape of the World*

      IISS Lead Speaker: Dr. Patrick Cronin
      Indian Lead Speaker: Prof. Ummu Salma Bava

      IISS Discussant: Sir Michael Quinlan
      Indian Discussant: Dr. Manpreet Sethi

1300 hrs.   : Lunch

1400 hrs.   : Session II–*International Terrorism*

      Indian Lead Speaker: Mr. B. Raman
      IISS Lead Speaker: Mr. Nigel Inkster

      Indian Discussant: Dr. Navnita Chadha Behera
      IISS Discussant: Sir Hillary Synnott

1530 hrs.   : Tea Break

1545 hrs.   : Session III–*Energy Security*

> IISS Lead Speaker: Col. (Retd.) John Gill
> Indian Lead Speaker: Dr. (Ms) Shebonti Ray Dadwal
>
> IISS Discussant: Mr. Rahul Roy-Chaudhury
> Indian Discussant: Dr. J. Nanda Kumar

1730 hrs.   : Programme Ends

Appendix 2

MEA–IISS Foreign Policy Dialogue
IISS, London, 5 February 2007

PROGRAMME

From 09:30 hrs   : MEA delegation & invited participants arrive

10:00–10:05 hrs   : Welcome Remarks by Dr. Patrick Cronin, IISS
                    Director of Studies

10:05–10:40 hrs   : Opening Addresses

> Keynote Address by H.E. Ambassador
> Arif S. Khan
> Additional Secretary (Public Diplomacy),
> Ministry of External Affairs and Address by
> Mr. Adam Thomson Director, South Asia and
> Afghanistan Foreign & Commonwealth Office

10:40–11:00 hrs   : Discussion

11:00–11:20 hrs   : Tea/Coffee Break

11:20–12:50 hrs : SESSION I: Growing Asian Economies: Challenges and Opportunities

Chair: Sir Hilary Synnott, IISS Consulting Senior Fellow for South Asia and the Gulf

MEA/India Speaker: Professor Pulin B. Nayak, Director, Delhi School of Economics
IISS/UK Speaker: Mr. Bill Emmott, former Editor, *The Economist*

(Examined the unprecedented simultaneous rise of two economies in Asia with a focus on global energy and environmental issues; India's infrastructure, agricultural and manufacturing challenges and opportunities, and prospects for sustained economic growth.)

12:00–12:50 hrs : Discussion

12:50–14:00 hrs : Sandwich Lunch

14:00–15:30 hrs : SESSION II: Terrorism & Transnational Security Challenges in South Asia

Chair: Rahul Roy-Chaudhury, IISS Research Fellow for South Asia

MEA/India Speaker: Dr. C. Raja Mohan, Strategic Affairs Editor, *Indian Express*
IISS/UK Speaker: Sir Hilary Synnott, IISS Consulting Senior Fellow for South Asia and the Gulf

(Examined security challenges and threats including terrorism in the subcontinent.)

14:40–15:30 hrs : Discussion

15:30–15:50 hrs : Tea/Coffee Break

15:50–17:20 hrs : SESSION III: UK–India Partnership in a Global Perspective

Chair: Dr. Patrick Cronin, IISS Director of Studies

MEA/India Speaker: Ambassador M.K. Rasgotra, Convenor, National Security Advisory Board, Prime Minister's Office
IISS/UK Speaker: Professor James Manor, Institute of Development Studies, University of Sussex

(Examined challenges and prospects for UK–India cooperation, with special emphasis on economic and political issues.)

16:30–17:20 hrs : Discussion

17:20–17:30 hrs : Concluding Remarks

18:30–20:30 hrs : Reception and Dinner at IISS

Index

About the Editor and the Participants

Amit Dasgupta assumed charge as the Joint Secretary of the Public Diplomacy Division, Ministry of External Affairs, Government of India in January 2007. He joined the Indian Foreign Service in 1979 and has served in various capacities in Delhi and abroad. He has edited a number of books and is currently working on two books on India; his first fiction title *In the Land of the Blue Jasmine* has gone into its second edition.

The Public Diplomacy Division of the MEA was specially created in April 2006 to respond to the need for open and regular dialogue between foreign policy practitioners on the one hand, and civil society, non-governmental organizations, academia, think tanks, business and industry, and the media, on the other. It is believed that such a dialogue and interaction could help shape public opinion and contribute towards a more informed understanding of India's foreign policy interests and concerns. Simultaneously, since the dialogue was essentially meant to be a two-way process, useful suggestions and inputs were expected to be received from the two different dialogue partners and stakeholders, which would assist in foreign policy formulation and implementation. The activities of the Public Diplomacy division are situated both in India and abroad.

The outreach activities of the Division are implemented in various ways and include organizing seminars and workshops, holding interactive discussions and brainstorming sessions, organizing an incoming and outgoing visitor's programme, commissioning documentary films and books, publishing the *India Perspectives* magazine, etc.

The International Institute for Strategic Studies (IISS) is amongst the world's leading think tanks on international security, political risk and military conflict. The IISS provides trusted and independent analysis for professionals and institutions, cutting-edge information on global developments and their effect on political and economic affairs. Based in London, it has offices in the US and Singapore.

The IISS was founded in 1958 in the UK by a number of individuals interested in how to maintain civilized international relations in the nuclear age. Much of the institute's early work focused on nuclear deterrence and arms control and was hugely influential in setting the intellectual structures for managing the Cold War. The institute grew dramatically during the 1980s and 1990s, expanding both because of the nature of its works and its geographical scope. Its mandate became to look at the problems of conflict, however caused, that might have an important military content.

The IISS is the primary source of accurate, objective information on international strategic issues for politicians and diplomats, foreign affairs analysts, international business, economists, the military, defence commentators, journalists, academics and the informed public. The institute owes no allegiance to any government, or to any political or other organization.

The institute's high-profile publications are both timely and authoritative. Among the titles are the annual *The Military Balance*, an inventory of the world's armed forces; *Strategic Survey*, an annual retrospective of the year's political and military trends; the *Adelphi Paper* monograph series, which provides in-depth analysis of general strategic issues; *Survival*, a quarterly international relations journal; and *Strategic Comments*, containing short briefings on breaking strategic issues.

The institute's conference activities—such as the annual Ministerial *Shangri-La* and *Manama* Dialogues in Singapore and Bahrain—are considered to be at the forefront of public policy development, especially given that its convening power is such that it can often bring government officials and others together in formats and circumstances that they could not easily manage for themselves.

The institute's staff and governing boards are international and its network of some 3,000 individual members and 500 corporate and institutional members is drawn from more than 100 countries.

The institute's South Asia programme—launched in October 2003—publishes research and assessments of key regional political and security issues and evaluates policy options. It also provides para-diplomatic support by organizing meetings and conferences for senior South Asian officials and experts to discuss regional and international security issues.

Ummu Salma Bava is Professor for European Studies and Coordinator of the The Netherlands Prime Minister's Grant in the Centre for European Studies, School of International Studies, Jawaharlal Nehru University (JNU), New Delhi. She joined JNU in April 2004 and prior to that, she taught at the University of Delhi (Zakir Husain College) from 1988 to 2004. Her expertise on Indian foreign policy has been acknowledged and she is one of the few to be recognized as an Associate Fellow of the Asia Society, New York. She has a doctorate in West European Studies. She was research scholar at the Free University, Berlin and also studied at the University of Vienna and Uppsala University, Sweden.

Ummu Salma Bava's areas of interest are Indian Politics and Foreign Policy, India–EU Relations, European and South Asian Security and Politics, Regional Integration and Organization—European Union, NATO, SAARC and Asia–Pacific—EU Foreign and Security Policy, International Politics/Globalization/Norms, German Politics and Foreign Policy, Transatlantic Relations and Conflict resolution. She is the author of one book and has several published articles/papers to her credit.

Navnita Chadha Behera is a Professor at the Nelson Mandela Centre for Peace and Conflict Resolution, Jamia Millia Islamia. She has earlier served as a Reader at Delhi University, as an Assistant Research Professor at the Centre for Policy Research, Delhi, and as Assistant Director, Women in Security, Conflict Management and Peace (WISCOMP). She was a Visiting Fellow at The Brookings Institution (2001–2002) and at the University of Illinois at Urbana–Champaign (1997–1998).

Navnita Chadha Behera has published widely in India and abroad. She has several books to her credit: *Demystifying Kashmir* (2006, 2007); *Facing Global Environment Change and Globalization: Re-conceptualizing Security in the 21st Century* (2007); *Gender, Conflict and Migration* (2006); *State, Identity and Violence: Jammu, Kashmir and Ladakh* (2000); *State, People and Security: The South Asian Context* (2001); *Perspectives on South Asia* (2000); *People-to-People Dialogues in South Asia* (2000); *Beyond Boundaries: A Report on the State of Non-Official Dialogues on Peace, Security and Co-operation in South Asia* (1997) and *International Relations in South Asia: Search for an Alternative Paradigm* (2008).

Patrick Cronin is the Director of Studies at the IISS, London. He joined the IISS in September 2005 to direct the institute's wide-ranging global research programme.

Patrick Cronin was Senior Vice President and Director of Studies at the Centre for Strategic and International Studies (CSIS) in Washington from 2003 to 2005. Earlier, he was the third ranking official at the US Agency for International Development (USAID), and also in charge of setting up a new development agency, the Millennium Challenge Corporation, for the White House. He also served as Director of Research and Studies at the US Institute of Peace, and held various positions at the National Defence University's Institute for National Strategic Studies, including Head of Asian Studies and Deputy Director of the Institute. He received both his doctorate and master's degrees at the University of Oxford, in 1984 and 1982, respectively.

Shebonti Ray Dadwal is currently a Research Fellow at the Institute for Defence Studies and Analyses (IDSA), New Delhi, specializing in energy-related issues. She has presented several papers in national and international seminars and conferences and has several peer-reviewed articles, projects and policy papers on issues related to energy security to her credit. She is also the author of a book, *Rethinking Energy Security in India* (2002).

Prior to joining IDSA, she worked with *The Financial Express* as Senior Editor and has also served as Deputy Secretary at the National Security Council Secretariat, New Delhi.

Colonel (Retired) John Gill is Senior Visiting Fellow at the IISS, London. A Professor on the faculty of the Near East–South Asia Centre in Washington DC, he retired as a Colonel after serving for more than 27 years as US Army South Asia Foreign Officer. He has been following South Asia issues from the intelligence and policy perspectives since the mid-1980s with the US Joint Staff, the US Pacific Command staff, and the Defence Intelligence Agency.

Colonel Gill served as Military Advisor to Ambassdor James Dobbins, the US envoy to the Afghan opposition forces (2001–2002). From August 2003 to January 2004, he served in Islamabad as the liaison officer to the Pakistan Army for the US forces in Afghanistan. His publications include *Atlas of the 1971 India–Pakistan War*, chapters in *Strategic Asia* (2003 and 2005), and a chapter in *US–Indian Strategic Cooperation into the 21st Century: More than Words* (2006).

Nigel Inkster is Director of Transnational Threats and Political Risk at the IISS, London. From 1975 to 2006, he served in the British Secret Intelligence Services (SIS). He was posted in Asia, Latin America and Europe and worked extensively on trans-national issues.

Nigel Inkster spent seven years on the Board of SIS, the last two as Assistant Chief and Director for Operations and Intelligence. He is a Chinese speaker and graduated in Oriental Studies from St. John's College, Oxford.

Arif S. Khan joined the Indian Foreign Service in 1975 and has served in various capacities both in India and abroad; he is presently India's Ambassador to Rome.

J. Nanda Kumar is a Researcher at the Institute for Defence Studies and Analyses (IDSA), Delhi, specializing on 'Energy Security Issues'. He has been working on issues related to India's energy security, China's energy geopolitics, non-state threats to energy security, oil and gas geopolitics in the Asia-Pacific region, and alternative and renewable energy sources. He is also a Research Associate with the Indian Institute of Public Administration, New Delhi and a Visiting Research Fellow at the Japan Institute of International Affairs, Tokyo.

J. Nanda Kumar has several published articles in Indian and international journals to his credit.

Shivshankar Menon assumed charge as the Foreign Secretary of India in October 2006.

Sir Michael Quinlan is Consulting Senior Fellow at the IISS, London. He spent his main career in the UK Civil Service, primarily in the defence field. He worked extensively on the nuclear-weapon policy and arms control in both the national and the NATO context. His final post was as Permanent Under-Secretary of State in the UK Ministry of Defence. For seven years thereafter he was Director of The Ditchley Foundation, which runs a high-level programme of international conferences. Sir Michael was educated at the University of Oxford.

Sir Michael has written and lectured extensively on International Security issues and in 1997, the Royal United Services Institute for Defence Studies published his monograph 'Thinking About Nuclear Weapons'. Most recently, he published 'India–Pakistan Deterrence Revisited' in the Autumn 2005 edition of IISS' *Survival*. He is a Visiting Professor in the Department of War Studies at King's College London.

B. Raman joined the Indian Police Service in 1961. He served in Madhya Pradesh as a police officer from November 1962 to July 1967. He joined the External Intelligence Division of the Intelligence Bureau of the Government of India in July 1967, and moved over to the Research and Analysis Wing (R&AW) of the Cabinet Secretariat, Government of India. He served as the head of the Counter-Terrorism Division of the R&AW between 1988 and 1994. He retired as Additional Secretary, Cabinet Secretariat, Government of India, on 31 August 1994. He is presently the Director of the Institute for Topical Studies, Chennai.

B. Raman was a member of the Special Task Force appointed by the Government of India in 2000 for revamping the intelligence apparatus. He was a member of the National Security Advisory Board (NSAB) of the Government of India from July 2000 to December 2002. He was a member of the Working Group on Terrorism of the Council for Security Co-operation in the Asia Pacific (CSCAP) in 2002 and 2005. He testified on terrorism before the US House Committee on Armed Services in June 2002 and a sub-committee of the US House Committee on International Relations in October 2003. Raman is the author of three books: *Intelligence: Past, Present & Future, A Terrorist State as a Frontline Ally* and The *Kaoboys of R&AW: Down Memory Lane.*

Rahul Roy-Chaudhury was a Senior Fellow for South Asia at the IISS, London. He heads the institute's South Asia programme, which publishes assessments on key regional political and security issues and organizes high-level meetings and conferences.

Earlier, Rahul Roy-Chaudhury was a Senior Research Fellow at the Department of War Studies at King's College London, and served in the National Security Council Secretariat in the Prime Minister's Office in India. Prior to his official appointment, he was on the faculty of the Institute for Defence Studies and Analyses (IDSA) in New Delhi. He has published two books, *India's Maritime Security* (2000) and *Sea Power and Indian Security* (1995). He was educated at the University of East Anglia and University of Oxford.

Manpreet Sethi is Senior Research Fellow, Centre for Air Power Studies, New Delhi where she is engaged in a research project on 'India's Nuclear Strategy'. She heads the Nuclear Security division at the Centre.

Manpreet Sethi received her Ph.D from the Latin American division of the School of International Studies, Jawaharlal Nehru University, New Delhi in 1997 and thereafter served on the research faculty of the Institute for Defence Studies and Analyses (IDSA), New Delhi for a number of years where she conducted research on issues related to nuclear proliferation and disarmament. She worked for the Centre for Strategic and International Studies from 2002 to 2006 as Senior Fellow. She is the author of the book *Argentina's Nuclear Policy* (1999) and co-author of *Nuclear Deterrence and Diplomacy* (2004). She also has various published research papers and newspaper articles to her credit.

Sir Hillary Synnott is Consulting Senior Fellow for South Asia and the Gulf at the IISS, London. He was the Coalition Provisional

Authority's Regional Coordinator for Southern Iraq from July 2003 to January 2004 and the British High Commissioner in India from 1993 to 1996. From 1996 to 1998, he was Director for South and South East Asian affairs at the Foreign and Commonwealth Office in London. He also served in Amman, Paris and Bonn. Before joining the British Diplomatic Service, Sir Hillary spent 11 years in the Royal Navy where he was a submariner.

In 1999, Sir Hillary published IISS' *Adelphi Paper 322*, 'The Causes and Consequences of South Asia's Nuclear Tests', and, most recently, 'State-Building in Southern Irag' in the Summer 2005 edition of IISS' *Survival*. Sir Hillary was educated at the Cambridge University.